Widow Woman's Ranch

& other stories

Memoirs of a Country Vet

-2-

David E. Larsen, DVM

Wiley Creek
Publications

Widow Woman's Ranch & other stories
Wiley Creek Publications
©2021 David E. Larsen, DVM
All Rights Reserved

David E. Larsen, DVM
PO Box 117
Sweet Home, Oregon 97386
email: d.e.larsen.dvm@peak.org
website: davidelarsendvm.com
blog: docsmemoirs.com

IISBN: 978-1-7367484-6-6

Book design, editing & production:
David E. Larsen, DVM/Wiley Creek Publications

Cover design: Eva Long/Long On Books
Cover composite photos: Idella Maeland/Unsplash
                         Kerry Snelson/Adobe Stock

Printed in the U.S.A.

*To my patients, and their owners
from the last forty-some years.
All of whom have made these stories possible.*

CONTENTS

Author's Note • 1
Hunting Dogs for Sale • 3
Widow Woman's Ranch • 10
On a Thanksgiving Eve • 16
Edith and Coco • 23
Under the Old Plum Tree • 28
Can We Eat Her? • 31
Old Three Toes • 36
Too Many Legs • 39
Cookie's Litter • 44
Long Road Home for Tramp • 49
One More Pregnancy • 53
Polyradiculoneuritis (Coonhound Paralysis) • 60
A Perfect Delivery • 66
The Shock of It All • 71
Ali • 75
The Upgrade • 81
Granny's Instructions • 85
Peanut Digger • 89
Toby's Sore Eye • 93
Wild Horses • 97
Uterine Twist, Which Way Do We Turn? • 102
One Bite Deserves Another • 106
The Perfect Shot • 110
Another Witch, Another February • 116
A Bear in the Backyard • 120
All the Better to See You • 124

The Thomas Splint • 128
The Turpentine Compress • 133
A Hasty Exam • 138
A Lesson Well Learned • 141
A Surprise Visit • 147
Agroceryosis, The Lack of Groceries • 152
Back to Her Old Self • 157
Benefits of Experience • 162
Rosebud's Wire • 165
Choose Your Surgeon with Care • 170
Colleagues • 175
Don't Put Her in the Barn • 180
All Hell Broke Loose • 184
Driving Blind •189
Elk Delivery • 193
Fleeing the Flea • 197
Harry and Buffy • 201
The Berserk Mule • 206
Hot Tub Skin Infection • 210
Ruth and the Goose • 215
Cows Never Eat the Stuff • 219
Don't Die on Me Now • 223
Meat is Life • 227
Run for Your Life • 230
Acknowledgements • 234
Photo Credits • 235
About the Author • 237

Author's Note

These memoirs are gleaned mostly from my memory, as few early records survive. They are presented in a rough chronological order. But in a small town mixed veterinary practice in the 1970s and 1980s, there was little control in what came through the door. And to some degree, these memoirs try to reflect that chaos.

This is the second book in a series of hopefully four or five books. Books will be similar to this book with short stories of specific snapshots in my life. In the third book I will include some stories of my early life and my Army experience. Those stories will provide a little insight into the making of a veterinarian.

Hunting Dogs for Sale

Cascadia is a small community located up the river from Sweet Home. It is currently little more than a wide spot in the road; however, there is a state park and a small church. And there is a small collection of houses and small farms scattered across the open spots in the forest.

In our early years, when there was still a lot of logging activity in the national forest, Cascadia was a thriving little community. It had an elementary school, a store, a church, a post office and the state park.

Cascadia was also a center for the illegal marijuana trade in the area and probably the state. Except for Mountain House, there was nothing much except mountains and timber east of Cascadia until you came to Sisters, some seventy miles over the Cascade Mountains.

I had several clients from Cascadia who habitually carried overdue accounts for most of the year. They would come in, usually in October, with a large roll of hundred dollar bills, peel off the required amount, and thank us for being understanding for their late payment. Occasionally, especially when they had a big harvest, they would pay an extra couple of hundred dollars toward the coming dry period.

That was the background when Doug came in with a sick puppy. This was during the early days of the parvo virus epidemic sweeping the state and the country. Parvo is a devastating disease for young puppies, and this was before we had a vaccine for the disease. Mortality rates were high, and treatments were not yet standardized. Treatment was expensive for the standards of the day.

Doug was a middle-aged guy with rugged features. Short in stature, he walked a little bent over. He had worked in the woods in his early years, but back injuries had put him on disability. His hair on top was thinning, but his handshake remained firm. I am sure he had some other income because he was one of the annual accounts.

This scrawny pup was approaching twenty pounds, had short grey hair, and was sick enough that he wanted to lie down during the exam. Joey was in the clinic record as a mixed breed. He probably had some pit bull in him, but that was just a guess.

"He's been vomiting for a couple of days, and this morning we noticed some bloody diarrhea," Doug said as I was starting to examine the pup.

I lifted the skin on the back of his neck, probably over 10% dehydrated, pale membranes, and quite depressed. The rectal temperature was depressed, 99.0°, and there was blood dripping off the thermometer. I had not seen too many cases of parvo virus in Sweet Home, but this was the likely diagnosis. We had no rapid means of making that diagnosis at the time.

"Doug, this pup has parvo. Parvovirus is a new disease going around, and it kills a lot of pups. It is very contagious. We will need to keep this guy in isolation. Treatment can be expensive, and I can't carry that kind of an expense on your account until next fall."

"Doc, I've never heard of parvo. Are you sure?"

"We can get a diagnosis by sending some samples to the diagnostic lab, but this pup is going to be cured or dead by the time we get results. If we do some blood work, we can almost confirm the diagnosis, and if the white blood cell numbers are too low, that will tell us that his chances of recovery are not good. Right now, we see as many as ninety percent of dogs die. I think my rates of recovery are higher than that, but the numbers that I have seen are few."

"I can't afford a bunch of lab work. My problem is bigger than just this pup," Doug confessed, hanging his head a little to avoid eye contact. "I have thirteen pups at home, and this morning, about half of them are vomiting. I was hoping you could give me something for all of them."

"Thirteen pups, are they all from the same litter?"

"No, they are from a couple of litters. But they are all about the same age and size. These are valuable pups, Doc. I know they don't look like much."

I looked at the pup. Valuable indeed, I thought. I was going to have trouble getting paid for this pup, let alone for thirteen puppies. Doug's wife and daughter could maybe give injections and subcutaneous fluids. With a bit of luck, that might be all they would need. If we saved half of them, we would be doing well.

"So Doc, do you think you could bring a bunch of medicine up to the house and show me what to do? Then I could treat them at home. That would save me a bunch of money, and we might get lucky. I can't afford to pay a big fee right now, but I could afford a house call and the medicine."

"You have to understand, you could lose the whole bunch with that plan," I explained. "I can bring some fluids and some injectable antibiotics up to your place and show you how to use it. But if we save half these pups, we are going to be doing well.

"Yes, I understand the risk, but that is just the way it is going to have to be right now. When I sell a couple of these, I will have plenty of money to pay for the medication and the visit."

There was the key: "When I sell a couple." This was going to be a credit, and he is planning to sell a couple of fifty-dollar pups to pay the bill.

"Doug, we could also lose the whole bunch. I am going to have to have some money to cover the drug cost, at least."

"I have a hundred dollars tucked away still from the fall harvest I could give you. Would that help?"

"Okay, I will do a quick blood count and a fecal exam. We want to make sure this isn't a bad case of salmon poisoning. You could have had someone throw a fish into your yard if they were tired of listening to the pups. You go ahead and take this pup home, and I will be up there right after lunch. You might refresh directions with Sandy on your way out the door."

The fecal exam was negative for salmon disease, only a few roundworms. Doug probably wormed these pups with chewing tobacco. The white blood count was low but pretty good for a puppy with parvo virus. We might have a chance with this plan.

I loaded the truck with 3 cases of fluids, Ringers Lactate, three dozen IV administration sets, half a box of sixteen gauge, one and a half-inch needles, and a box of three cc, twenty-two gauge syringes. I mixed a bottle of ampicillin and grabbed a bottle of gentamicin.

The drive up the river will be a pleasant one. There is seldom much traffic other than a couple of logging trucks, and the river should be running clear and full for early June. It is only 19 miles to Doug's place, but the drive will take well over a half hour.

The bushy regrowth on a couple of maple stumps mostly obscured the turnoff to Doug's house. Several houses shared the long gravel driveway. These houses were all the same. Small, unkept yards and untrimmed hedges made finding numbers difficult.

Numerous marijuana plants were evident in the ditch along the driveway. Unattended, but they grew well in our moist climate. It reminded me of the "plant a pot seed" drive conducted by the hippie crowd when I was in vet school in Colorado. They had pot plants growing all over town, drove the sheriff and police nuts.

Doug's was a small house, probably built in the thirties and added onto several times. The yard was overgrown, and a large clump of marijuana plants grew in the corner of the yard. Pups were everywhere in the yard. This yard would be the local infection source for the parvo virus for the next year or so.

Doug and his daughter, Debbie, were at the gate to greet me. Debbie was older than I had expected. She looked in her early thirties, dishwater blond, slender, with an attractive figure. We unloaded everything, and we grabbed a pup to demonstrate how I wanted them to treat them.

"How do we know which ones to treat?" Doug asked.

"I think it would be a good idea to treat them all. I'll do a quick exam on each pup. But any pup not sick today will be sick tomorrow."

I showed Doug's daughter how to give the injections and how to administer the fluids.

"Give each pup about 500 cc fluids under their skin on their back, between the shoulder blades is easiest. It will make a big lump, but that will go down quickly as the fluid is absorbed. You may notice some swelling around their elbows, but don't worry about that."

She took the instruction in stride and already had a chart to keep track of what she would give each pup.

"You give me a call every morning at about ten. I should be done with surgeries by then, and I'll have time to talk with you. We could lose some of these guys. We are going to need to get lucky," I said as I was getting ready to leave.

Doug came out and handed me a hundred-dollar bill. "This is all I have right now, but all I need to do is sell a couple of these pups, and I can make things right with you, Doc."

You might need to sell more than a couple of fifty-dollar pups, I thought to myself. "Give me a call in the morning so we can keep track of how we are doing. And Doug, hold your mouth just right when you sleep. We are going to need all the luck we can muster."

The first morning the call came right at 10:00. The pups were still alive, but all of them were vomiting. Debbie had treated them all with no problems. The following day everything was going fine, and Debbie thought most of the pups were feeling better. By the third day, it looked like we were out of the woods with only one pup still not feeling well. It looked like we were going to get through this smelling like roses. This was probably something other than parvo.

Now all I needed was for Doug to pay the bill.

***

It was a month later when Doug came into the clinic. He had a broad smile on his face as he pulled a roll of bills from his pocket. "You did great, Doc. Those pups all came through that parvo without a problem. I've already sold five of them."

"That's good. We got lucky. We have a new vaccine coming out. You just have to remember to vaccinate your dogs and all the pups next time. That virus stays in the ground for over a year."

"I have enough to pay the bill, and I can pick up vaccine for all the dogs. It has been a while since I wormed them, except for a tobacco chew, so I better pick up some worm pills also."

"Are you sure you have enough for the bill and all of the other stuff? You only sold five pups."

"These are valuable pups, Doc. I sell them as hunting dogs."

"Hunting dogs! What do you mean by that, Doug?"

"I train them up in the mountains, and I sell them all over the country. I don't have any problems selling them. If I do, I just lower the price to five hundred dollars, and they are gone almost overnight."

Five hundred dollars was unheard of for dogs at the time. Even the best purebred dog in Sweet Home would not sell for five hundred dollars.

"Doug, that is unreal. What is your initial asking price?"

"I sell most of them for seven hundred and fifty dollars, plus whatever I need to add on to cover the shipping cost. I sell these pups all over the country. I just run a couple of little magazine ads."

"What kind of training do you give them? I mean, what can they hunt that makes them that valuable? Guys that run purebred hounds don't get that kind of money for their pups."

"We train them up in the mountains. There is a lot of stuff that goes on in those woods. People never see a lot of the stuff unless they spend a bunch of time out there."

"What are you talking about?" I ask, somewhat afraid of the answer.

"Doc, where else can you buy a hunting dog that has actually been trained to track Sasquatch?"

Widow Woman's Ranch

I could see Dan waiting at the gate to the barnyard when I pulled into his driveway off of Pleasant Valley Road. There was still some snow on the ground from a late spring snowfall, and it added a chill to the air.

"Dan, I got the message that you wanted me to stop by, but I didn't get any other information," I said as I extended my hand.

Dan shook my hand. His hands were heavily calloused, and his fingers bent from arthritis. I am sure his handshake was much firmer in past years.

"I want you to look at my old horse, Joe," Dan said. "This time of the year, I keep him in the barn. I've noticed that he has one heck of a time eating. He takes a mouthful of grain, and more of it dribbles back into the feed rack than he swallows."

"That might be something pretty simple," I said. "How long has it been since his teeth have been floated?"

"They probably have never been floated since I don't know what that means."

"The horse's teeth continue to erupt throughout their life. They wear against themselves, and sometimes, when they get a few years on their mouth, they develop sharp points on the edges of the teeth that need to be filed off."

"You just say, open wide, I guess," Dan said.

"Some horses object to the procedure more than others, but we have a little device to help hold the mouth open. Some horses stand right there and let it happen, some need a twitch, and then there are a few who need some drugs to help them relax."

"Joe, he's a pretty mellow old horse. He's sort of like Mom and me. He was a lot prettier twenty years ago than he is today."

"Let's go get a look at Joe," I said.

"I'm a little embarrassed to take you into the barn," Dan said. "This place looks like a widow woman's ranch anymore. I'm too damn old to keep it up. Our son, Stan, died in that war, and our daughter doesn't live close. She tries to help some. But you probably know how it is when you are working and have a young family. There is just so much time you have to give to the old man."

"I'm sorry about your son. There were far too many young men lost over there. I was in the Army for four years, but I was able to avoid Vietnam. I had a good friend who came home in a box, though. I'm glad that it's over."

Dan didn't respond to my comment. He busied himself with the gate that we had stopped at on the driveway leading to the barn. Dan was having some trouble untying some baling twine that held the gate closed.

"I've never heard the term widow woman's ranch," I said.

"Nothing is fixed. All the fences lean this way or that. Everything is held together by baling wire or twine. The wire lasts a lot longer than the twine, but they don't bale hay with wire much anymore. Or maybe, I just don't buy alfalfa much anymore. Twenty years ago, I would've replaced any leaning post. Or at least, reset it. Now I just support it with a mesh made out of twine. It's a good thing I don't have much stock anymore. We feed out a steer for meat, for ourselves, and our daughter's family. And then there's Joe. And Joe knows he doesn't want to get out. He's got it made here, three square meals a day, and nothing is expected in return. Even the grandkids don't seem to want to ride him anymore."

I helped Dan with the large barn door. We had to lift it a bit, and then it would slide. It looked like the rollers needed a little grease, but I wasn't going to say anything.

I was shocked at the inside of the barn. It was immaculate, like stepping into a barn twenty years in the past.

"I try to keep this place like Stan would remember it," Dan said.

Joe was in a large stall. He whinnied and tossed his head happily.

Joe was old, a buckskin. He was probably a striking horse in his day. Now his face was grayed, and his muscle mass was fading.

"Let's get a halter on him so I can look at that mouth," I said as I reached for a halter and lead rope hanging on a hook at the gate leading into the stall.

"Don't use that one," Dan said. "Stan hung that one there the last time he rode Joe before going to Vietnam. Joe was Stan's horse, you see. Joe is the only connection I have to Stan. I worry what will happen to him if this old guy outlives me."

I could see some moisture in Dan's eyes as he spoke. I had to look away for a moment and take a couple of deep breaths before I tried to talk.

"I'm sure your daughter will take care of Joe," I said.

"She has nowhere to keep him, but I guess she could find someone to take care of him. I have it all spelled out. I have a place picked out for him behind the barn. Joe and I go out at times and talk about how things used to be when Stan was around."

I grabbed the old halter that Dan had been holding and stepped into the stall. I needed to get to work to change the subject.

Joe nuzzled me when I slipped the halter over his nose. I tied the lead to a ring hanging on the feed rack. Joe had no problem when I ran my index finger along the insides of his cheeks to feel the points on his back teeth.

I grabbed Joe's tongue and pulled it to the side, causing Joe to open his mouth a little. With a small penlight, I got a good view of the left side of his mouth. Switching hands and pulling the tongue to the other side, I viewed the right side of his mouth.

Joe had jagged points on the inside of his lower cheek teeth and the outside of his upper teeth. I could see sores on both sides of his tongue and on the inside of both cheeks. Joe should feel much better with these teeth floated.

Joe was remarkably tolerant. I grabbed his tongue and inserted the float in the left side of his mouth. With long slow strokes of the float blade, you could hear the points disappear as the sound went from a rough rasping sound to a smooth, almost silent sound. I finished the floating in a couple of minutes.

I smiled as I felt Joe's teeth after I was done.

"These are as smooth as can be now. I think you'll see a big difference for Joe."

"I hope so. I am a little surprised that Joe wasn't bothered by that whole thing," Dan said.

"Yes, he is pretty exceptional to stand there and take it with no restraint and no speculum."

"Do you think he is going to be able to eat now?"

"I think you'll find he is a new horse. But to be sure, I'll check with you in a few days."

***

I was driving by Dan's place a few days later, and when I noticed him coming out of the barn, I stopped to ask about Joe.

"Good morning, Dan," I said. "I just wanted to check on how Joe was eating after I worked on him the other day."

"He is doing great. Doesn't dribble a bit of grain. I think he enjoys eating now. If I had known that was a problem, I would've had you do his teeth a long time ago."

"At his age, we should plan to check him every year, just to keep him as comfortable as we can for his old age."

"That sounds like a good plan," Dan said. "I'll try to remember to give you a call."

***

It was probably three years later when Dan gave me a call. He wouldn't talk to Sandy when she answered the phone. He only wanted to speak with me.

"Doc, this is a terrible day for me," Dan said. "I think I need you to come and put Joe to sleep for me. He slipped going out of the barn the other day, and he must have hurt a hip or something. He can hardly walk."

"Do you want me to examine him first?" I asked. "It could be something simple that we could help with some medication."

"No, I think it is time for him to go see Stan. I've already had a neighbor over. He dug a hole with his backhoe, out behind the barn, under that big maple tree."

"When do you want to do this, Dan?" I asked.

"The sooner you can come, the better. Do you have time to come now?"

"I'll make time, Dan. I'll be out there in a few minutes."

Dan was waiting for me at the barn door. I helped him with the door again. Joe was lying down when we entered the barn. It was quite a struggle for him to get on his feet.

Dan had tears in his eyes and one running down his left cheek.

"Can you get Stan's halter down for me?" Dan asked. "This getting old stuff is no fun for me either."

I remembered that Dan didn't want to use this halter when I was out before to look at Joe.

"You want the one that Stan hung up?" I asked.

"Yes, we are going to bury him with it. I figure that is what Stan would have wanted."

I carefully, almost reverently, lifted the halter and lead rope off the hook. Dan took it and held it close to his chest as he walked through the gate to Joe's stall. Joe snickered softly as he smelled the halter. Dan slipped the halter onto Joe and patted him on the neck.

"We can go out the back door," Dan said as he started Joe toward the door. Joe hobbled along beside Dan. He barely touched his right hind hoof to the ground. Dan was correct. It was probably time.

Dan led Joe out to the big maple tree. Joe went to his knees with some coaxing and then flopped to his side with a big groan. The hole that was recently dug loomed large in my field of view.

"He knows the routine. I've been bringing him out on good days for the last year. He lays down, and I sit here with him, and we talk about the old days."

"Dan, do you want to stay for this?" I asked. "You could wait in the house or the barn."

"No, Joe wants me to be here. It's okay, Doc. Joe's going to see Stan."

"This is pretty fast stuff. I'm going to give him an injection to sedate him a little. Then, when I give him the big injection, he'll be gone in an instant."

"Okay, let's get it done."

Joe went quietly, resting his nose on Dan's legs when he was sedated.

Dan shed a couple of tears, patted Joe's neck, and stood up.

"Do you need me to help to put him in the hole?" I asked.

"No, the neighbor is going to come back with his backhoe. Our daughter will be here before too long. She and her husband will be able to take care of everything. You can stop at the house, and Sue will give you a check."

"There's no charge for this, Dan. Stan paid the bill some time ago."

I walked back to the truck alone, leaving Dan to pay his last respects to Joe. I sat in the truck for a moment, took a few deep breaths, and dried my eyes before pulling out onto the road.

On a Thanksgiving Eve

The barn was cold, but there was steam rising from the back of the young heifer. A dusting of snow on her back was melting fast. Bill and Connie Wolfenbarger had called with a heifer in labor. They were not regular large animal clients but did visit the clinic with their small dogs. I had been to their place several times to treat cows belonging to the Gilberts.

When they discovered a tail hanging from the heifer's vulva, they knew they had a problem. This meant the calf was in a true breech presentation and almost certainly dead. In a true breech position, the calf does not engage the cervix, and the cow doesn't go into hard labor. Most people will not notice a problem until the calf has been dead for a day or two.

Tomorrow was Thanksgiving, I would miss some of the prep for the family dinner. The evening snowfall was light but continuing. Hopefully, I could make it home before the roads became a problem. Sandy's folks were already at the house, so we didn't have to worry about anybody traveling tomorrow.

I tied the heifer's tail out of the way and started to wash her rear end. The hair on the tail came off with the slightest touch. I pulled on a plastic OB sleeve onto my left arm. With a little lube on my hand, I eased into her vulva to explore the birth canal. She strained hard when I reached the butt of the calf. No fluid was expelled with the strain. I pushed the rear of the calf with a couple of fingers. There was a spongy consistency under the skin and some crackling like I was popping air bubbles. The calf filled the entire birth canal, I could not advance my hand into the uterus, and I could not move the calf, it was wedged solidly into the birth canal. I withdrew my hand, the sleeve was covered with hair from the calf and the odor was slightly pungent.

"This calf has been dead for over a week," I said as I removed the sleeve. "It is emphysematous, blown up with gas. I'm not sure I am going to be able to get it out of her."

"What are our options?" Bill asked. I knew their daughter was a small animal veterinarian, maybe in California. I did not know her, but I would assume they would be a little more knowledgeable than most clients.

"Options are not many. The calf is in a breech position. That means its hind legs are retained and only the rump is presented. It is blown up so much that I cannot even insert my hand into the uterus. We are not going to be able to deliver this calf vaginally. I try not to do a C-section on a dead calf, but with all the emphysema I won't be able to do a fetotomy. That leaves us with two viable options. Option one is a C-section, which will be with risk and will be difficult."

"And the second option?" Bill asked.

"The second option is to get your rifle and shoot her now. It would not be fair to her to leave her in this situation," I said.

"Let's do the C-section," Bill said.

I double checked her halter to make sure she would not be choked if she went down. Then I changed the tail, tying it to the right side. I placed a rope around her neck with a bowline and ran it along her side and tied it to the alley fence, holding her against the fence. With her in a reasonably secure position, I clipped a wide area on her left flank, from her dorsal midline to the bottom of her flank.

I prepped her flank with Betadine Surgical Scrub. Then with 90 ccs of two percent lidocaine, I did a large inverted 'L' block of her left flank. I blocked a wider area than usual because I may need to make a longer incision than is usually required. This was not going to be an easy procedure. I repeated the prep after the block.

After laying out the surgical pack and supplies, I pulled on a pair of surgical gloves, more to keep my hands clean than to pretend that this was going to be a sterile procedure.

"We have a couple of major risks with this surgery," I explained as I prepared to make my incision. "The first one is that it is going be difficult to pull this uterus to the incision and second when I open the uterus, there is going to be no way to prevent the contamination of the incision and the abdomen. We are just going to have to depend on antibiotics."

Bill nodded, and I made a long incision down her flank, starting a few inches below her transverse processes and extending about 15 inches down her flank. The skin and subcutaneous tissues parted easily. Then I incised the muscles of the flank; they quivered as the blade divided them. When I incised the peritoneum, the abdominal content did not sink away from the incision with a characteristic rush of air into the abdomen. The distended uterus filled the entire abdomen. There was no trouble finding it or worry about moving the rumen to externalize the uterus. The abdomen was filled with the uterus.

I reached into the abdomen to the tip of the uterus. Cupping my hand around the tip of the uterine horn, I pulled. The uterus did not move. I tried to rock the uterus in the abdomen. Sometimes you could swing the uterus enough to make it easier to bring it to the incision. This uterus did not budge. Again and again, I tried to bring the uterus to the incision. I searched and found a hoof; I could not move the hoof.

I pulled my arm out, stretched and changed gloves. In this cold barn I was sweating profusely.

"Do you think I could give you a hand?" Bill asked.

"We might try that, if we could both get a hand under the end of the uterus, we might be able to make it move," I replied.

Bill stripped down to his waist and washed his hands and arms thoroughly. I stood on the head side of the incision and ran my left hand down to the tip of the uterus. Bill on the other side of the incision inserted his right arm. I guided his hand to the correct position. We pulled, we pushed, we tried almost every maneuver. The uterus did not budge.

Bill and I were almost nose to nose. Bill had sweat on his brow and sweat dripping off the tip of his nose. He looked me square in the eye.

"A woman couldn't do this," he said.

I smiled, "If you haven't noticed, I haven't got it done myself, yet."

We pulled out, and I rethought the situation.

"I am going to try one more thing," I explained. "I am going to open the uterus up here without externalizing it. I will then try to get hold of the calf's hoof and turn it up to the incision. The risks in doing this are many. I could spill content into the abdomen, I could tear the uterus, and even with a grip on a foot, I might not be able to budge this uterus."

"And then, if this fails, we are going back to option two. We will get your rifle and put this girl out of her misery."

That said, I incised the uterus in the middle of the flank incision. With a surgical glove and an OB sleeve on, I ran my right hand down the inside of the uterus. There was a front foot, I grabbed the leg just above the hoof and pulled as hard as I could. The uterus rolled and the hoof popped out of the incision. With my left hand, I incised the uterus over the hoof, and then I slipped an OB strap onto the hoof.

I handed the strap to Bill. "Keep that foot from going back into the abdomen."

Pulling my arm out of the upper incision, I enlarged the incision over the exposed hoof. Bill was able to pull the entire front leg out of the incision. I reached in and found the other leg. It came out quickly, and we attached it to the other end of the OB strap.

With both front legs out, I enlarged both the flank incision and the uterine incision. Now I was able to pull the head out of the incision. Then putting things down, I helped Bill put a hard pull on the calf. It was sort like pulling a basketball through a knothole but when the gas-filled abdomen of the calf finally cleared the incision both Bill and I almost fell as the rest of the calf followed with a swoosh.

The membranes and the calf landed on the barn floor in a splat. Then the odor hit us. Bill and Connie both gagged and had to turn back to the side door. When they opened that door, things were better, or maybe we were just adjusted. Bill grabbed the OB strap and pulled the calf out the barn door, and I returned my attention back to closing up this mess.

I washed and changed gloves. I put five grams of tetracycline powder into the uterus and pulled the open incision to the outside. This was a long incision in the uterus, and then there was the small incision higher on the horn. I elected to close the upper incision first, just in case the uterus would start to involute, and I would not be able to reach this incision. I closed it with a simple continuous suture.

The larger incision I closed with my standard Utrecht closure. After closing, I returned the uterus to the abdomen. I was concerned that there was probably a lot of leakage into the abdomen and pondered how best to deal with that event. There was no option to lavage the abdomen in the middle of this barn, so I just dumped another five grams of tetracycline powder into the abdomen.

I closed the flank incision with four layers. I used simple continuous in the peritoneum, interrupted mattress in the muscles and simple continuous in the subcutaneous tissues. I closed the skin with an interrupted mattress pattern. No need to spray for flies in this weather.

The heifer had to feel tremendous relief getting that mess out of her. She was going to need some additional antibiotics to keep her incision from falling apart. The easiest thing was to use some long-acting sulfa boluses. I gave her four boluses of Albon SR. That would give her five days of protection.

I untied her tail rope and the sideline. She was as calm as could be expected. I loosened the halter and slipped it over her head. She turned slowly and headed to the door, sniffing the floor a little as she passed.

"She should be okay for tomorrow, but I will check with you on Friday," I said to Bill as I was cleaning myself up.

"She will be just fine," Connie said. "Our daughter will be home for a week or two. She can check her tomorrow. We will let you know how she is doing. You go home and rest for dinner tomorrow."

At least I was going to have a few days to rest up with the holiday. I stopped at the clinic and cleaned all the equipment. It would be a real mess if I left it for the girls on Monday. I stripped down to the waist and washed thoroughly. The one mistake I made with the clinic was not putting in a shower. I thought I would wash here and go home for a shower. Then probably have to start working on dinner for tomorrow.

***

Friday morning, Bill called. "The heifer is doing great. Our daughter is impressed with how the incision looks. We told her the story, but I don't think she really believed us.

Two weeks later Bill called again to say they took the sutures out and the heifer continued to do well.

Edith and Coco

Edith was an older lady who you would see walking down the street in Sweet Home daily. I don't think she drove, but maybe she just preferred to walk. Her hair was always curled but not what you would call well-kept. Some would call her petite, and I am sure she was at one time, but I would say she was matronly petite. In any case, the crucial thing about Edith was she was always happy. We would notice that happiness in the clinic, and when you saw her walking, she always had a smile on her face. She could definitely enjoy the simple things in life.

For the first five years that we were in Sweet Home, Edith visited the clinic often. In fact, there was not a single month without a transaction on her account. This was unusual in that the average client might have three to five transactions per year.

She would come into the clinic with Coco always in tow. Coco was a mutt. Many people would call him an ugly mutt. His lower jaw protruded well past his upper jaw, and when you looked at him, he would often smile, but it looked like a snarl.

Coco's monthly trips to the clinic were more of a social event than a medical one. Coco was healthy as a horse, but there was always something that Edith wanted to be checked. I can't remember finding anything wrong with Coco.

One Saturday morning, we planned our day and hoped to take the kids to a movie in Albany when the phone rang. It was Edith, and she was sure Coco had a problem. My initial thinking was Coco never has a "real" problem. This would disrupt our entire weekend to go to the clinic and reassure Edith that everything was fine with Coco.

"Edith, are you sure this couldn't wait until Monday?"

But Edith persisted. "Doctor, I know there is somewhat dreadfully wrong. Coco is just not himself this morning!" she replied.

I was stuck, but it should only be a brief visit. I arranged to meet Edith at the clinic in fifteen minutes. Paul was home and could drive her and Coco to the clinic, so that would work.

Edith was smiling but concerned when she came through the clinic door. She thanked me profusely and reassured me that there was indeed something wrong with Coco. Coco groaned a little when I picked him up and put him on the table.

Maybe he has hurt his back, I thought to myself. His temperature was normal, but Coco was not wagging his tail and was not acting his usual happy self on the table. His heart and lungs were normal, and the oral exam was normal. Then I got to the abdominal palpation. Coco tensed his abdomen from discomfort. His bladder was distended and uncomfortable.

I had almost made Edith and Coco wait until Monday. And here we have Coco with a urinary tract obstruction. He could have been dead by Monday.

"Edith, when was the last time you saw Coco pee?" I asked.

"Well, he was outside this morning and lifted his leg several times, but nothing happened."

"Edith, Coco can't pee. Most likely, he has a stone blocking his urethra. If so, I will have to do surgery to remove the stone. I need to do some x-rays first to see if there are stones, how many stones, and where they are located. Most of the time, there is just one stone blocking the urethra, the tube from the bladder to the outside. The surgery involves opening the urethra and removing the stone. If there are more stones in the bladder, we will need to do abdominal surgery to remove them also." I explained.

"Surgery!" She exclaimed. "Shouldn't that wait until Monday?"

"No, we can't wait that long. Coco might be dead by Monday if we don't do surgery now. At the very least, he would have some major complications by then."

"You do whatever you need to do, Doctor," she said. "We have the money in the bank to pay for it, and we can't give Coco up."

"I will get some x-rays and call if anything changes in my thinking after the x-rays. Otherwise, I will call following surgery, and we will arrange to send Coco home sometime this weekend." I said.

We were going to have to get lucky to able to take the kids to a movie today. I took the x-rays, and sure enough, there was a stone stuck at the base of the os penis. The dog, like many animals, has a bone in his penis called the os penis. The urethra narrows slightly as it passes through a groove on the underside of the os penis. Most stones that cause obstruction are lodged in this location. Coco had no other stones visible in his bladder or elsewhere in his urethra. This would be an easy surgery.

I called Sandy and had her get the kids ready and come down to give me a hand. All the kids had observed many surgeries, so this would just be one more. I started getting Coco and the surgery suite ready, so we would be prepared to go the minute Sandy and the kids arrived. The plan was to do the surgery, recover Coco, and then run to the movie while Coco was resting in the kennel. We should be able to send him home when we return from Albany.

When Sandy arrived, we got started with the surgery. I induced anesthesia with IV pentothol and then put Coco on gas anesthesia. With him on his back, I clipped and prepped his posterior ventral abdomen. I could feel the stone. This should be a brief procedure.

I inserted an eight french urinary catheter. It came to a stop at the stone. I made a one-inch incision in the skin of the prepuce over the stone. Then I dissected through the soft tissues to the urethra. I pushed a forceps through the tissues on the dorsal surface of the penis to stabilize the area. Then with a careful incision, I opened the urethra over the stone. This incision was just long enough for me to grasp the stone with forceps and remove it. I immediately plugged the hole with finger pressure. I advanced the catheter into the bladder to empty it and avoid urine leakage into the surgery site.

After emptying the bladder, I left the catheter in place to ensure my closure did not narrow the urethra. I closed the urethra with interrupted 4-0 Maxon sutures. Then with the same suture material, I closed the subcutaneous tissue with a continuous suture pattern. Finally, I closed the skin with 4-0 nylon interrupted sutures. I infused a small amount of lidocaine for pain control and turned off the gas to start waking up Coco. Maybe fifteen minutes had elapsed. Since he would be unattended in a kennel after he was awake, I gave him fluid under his skin on his back rather than IV.

Recovery was pretty rapid, and Coco was up and about. He would be fine and should be able to go home when we got back from Albany. I gave Edith a call and reported favorable results. We arranged to meet her and Paul at the clinic when we returned from the movie.

The kids enjoyed the movie with Indiana Jones in Raiders of the Lost Ark. Sandy had tried to cover Derek's eyes in the spider scene.

But he was able to fight her off. The kids had been worried they would miss the movie because they had seen family plans set aside more than once by a phone call.

When we pulled up to the clinic, Edith and Paul were waiting out front in their car. They were talking and laughing and passing a whiskey bottle back and forth between them. Here was our happy little gray-haired lady who adored Coco, sitting in the car outside the clinic, drinking whiskey with her husband. Now my only concern was them driving home.

Under the Old Plum Tree

It was almost midnight in the early fall of 1977 when Lloyd called with a sick cow.

"The boys say she has been eating plums. She seems to be pretty sick, Doc. Do you think she will be OK till morning?" He asked, obviously quite worried about his favorite cow.

"How many plums do you think she ate, Lloyd?" I inquired, hoping I could justify rolling over to go back to sleep.

"The boys say she was under that tree all afternoon, and there are plums all over the ground. The limbs are hanging pretty heavy with them," Lloyd replies.

"It'll take me a little while, but I will be there shortly," I say as I throw my legs out of bed and start looking for my clothes.

Lloyd is a tall thin, soft-spoken man with a thick mustache and thinning hair. I have been to his ranch once before, but I have seen Lloyd and his dog at the clinic several times.

I turned onto Scott Mountain Road and passed Ayers' driveway. It's well past midnight as I took the curves going up Scott Mountain Road. It doesn't take long, and I break into the open fields of Pat's place. The moon is full, and the crisp autumn night is still. The stars are bright, and the sky is very striking out here, far removed from the city lights.

As I entered the timber starting down the backside of Scott Mountain, a bobcat suddenly was surprised by my headlights in the middle of the road. He ran helter-skelter ahead of me down the mountain road. With high banks on both sides of the road, he ran headlong, looking for an escape from the glare of my headlights. It's unusual to see a bobcat on the road, and I was a little surprised at the speed he is traveling. The road suddenly opened again, and he darted into the brush on the right side of the road.

Lloyd and his son are waiting for me at the door of a small barn next to the road. A very miserable cow is standing in the milking stall. She is not in a stanchion, which is probably a good thing because of the possibility of her falling.

The old Jersey, standing head down, was not wanting to move. The left side of her abdomen was quite distended with gas. I didn't need a rope or halter to handle her. She was miserable enough that she wasn't about to move.

I slide through the gate and start doing an exam. Her temperature is normal, and her chest sounds are normal. When I got to her belly, her rumen was distended and hardly moving. Then I start a rectal exam. Standing on her right side, I begin to insert my gloved left hand into her rectum. She has a major blowout of watery diarrhea, just missing me.

"That was close," I say. "Has she had that diarrhea for a while?"

"No, she has been fine until tonight," Lloyd said. "The boys say that she was eating plums all afternoon, out under the old plum tree we have out on the hillside. Most of the time, that tree doesn't have much fruit, but this year it's loaded."

Using a Frick speculum, a metal tube to keep the cow's teeth from chomping on the stomach tube, I passed my large rubber stomach tube into her rumen, the first stomach of cows. I blew a deep breath of air into the tube to clear it of obstructing rumen content. I pulled the tube from my mouth and pointed it away from me. The rumen gas filled the small shed. The smell of fermented plums was overwhelming. The old Jersey felt better with the relief of the gas. I pumped a gallon of mineral oil into the rumen. This is to aid in the passage of the plums through the gut. Then I pumped in a gallon of warm water with a pound of Carmalax powder dissolved in it. Carmalax is an antacid, laxative, and rumen stimulant all rolled into one.

"She'll be fine, Lloyd. But you probably don't want to be caught standing behind her in the morning," I say, smiling as I begin putting things away in the truck.

I enjoyed the night and the drive home; what a beautiful drive home with the full moon. I didn't see the bobcat again, even though I looked closely along the edge of the road where I had last seen him. He probably would not make the mistake of being caught on the road again for a long time. It was going to feel good to get back to a warm bed and snuggle close to Sandy.

***

When I called Lloyd the next day, the cow was doing great.

"I see what you meant last night," Lloyd said, "She sort of plastered the walls of the shed during the night."

Can We Eat Her?

"Bart, it's good to see you. Do you need to talk with Debbie?" I asked.

"If she is busy, that's okay. I can wait. I just wanted to tell her we would be gone by the time she gets home. We will have dinner ready for her and Lisa. She is just going to have to pop it in the oven."

"I'll send her out. They are just cleaning things up a bit," I said.

"While I'm here, Doc, I have a cow that is lying around a lot. She looks fine, but she has hardly moved in the last few days. I can't do it tonight, but is there some time in the late afternoon that I could get you to look at her?"

"I could get up there tomorrow. What time do you get home?"

"I can make sure that I'm home by three. It is easy for me to skip the last load of logs on Friday. Do you know where we are located?"

"I know you are on Whiskey Butte. But I can solve the problem. I will just have Debbie lead the way. I can make it an end-of-the-day call. She will like to leave the clean-up for the others."

"She will like that. She really appreciates this job, Doc."

***

It was close to four by the time we pulled into the pasture with the cow. It was almost a half-mile up the road from the house. The cow, an older Hereford, was lying down when we approached. She stood up but was reluctant to walk away.

I was able to examine her with no restraint. She had a moderately elevated temperature, but otherwise, the exam was pretty unremarkable. I palpated her ventral abdomen with some deep pushes, checking for pain from a wire. There was no response. Putting downward pressure at the middle of her back caused a definite groan. I waited a moment and then repeated the maneuver. She swung her head at me this time to emphasize her discomfort.

"Her back is pretty painful," I said. "It is hard to say what she did. It could be just soft tissue stuff, or she could have broken something."

"What can we do about it?" Bart asked.

"We aren't going to do an x-ray. I can give her a dose of Banamine and see if that helps."

"What does that do?"

"It is like a big dose of ibuprofen, just an anti-inflammatory medication. If that doesn't do it, I don't know. She might be salvageable if that temperature goes down."

***

I was trying to close down the clinic a little early on Saturday morning when the phone rang.

"Doc, this is Bart. That cow is down and can't get up. What do you think?"

"I am not sure there is anything more I can do for her." I said. "But I can run up and get a quick look at her. We are slow here this morning."

"I don't want to impose on your free time if there isn't anything to be done."

"That's fine, Bart. We don't have any plans for the afternoon, and it won't take me much time to get a quick look."

"Why don't you take your time and have lunch. Then bring Sandy and the kids up for the afternoon. We can barbecue dinner, and the kids can swim in the pool if the sun stays out. We haven't turned the heater on yet this spring."

"Okay, but this might not turn out very favorably for the cow. I don't want to ruin your afternoon with bad news."

"I'm a big boy, Doc. I have shot a cow or two before. If that is what we have to do, I can deal with it."

***

The cow was down. With a slap on the rear, she wouldn't even try to stand. The temperature was improved, just slightly above average.

"I think she has had it, Bart," I said. "She is pretty painful. It might be best to get your rifle."

"Do you think we could eat her?" Bart asked.

"Eat her? My first answer is no, she can't stand and has a temperature plus some Banamine on board. But, I guess it depends on how hungry you are. The meat isn't going to kill you. It is just not going to be very good. She has been stressed, down, probably has a significant injury. All of that is going to influence the flavor of the meat. Sort of like eating a gut-shot deer that took you a day to find."

"You know, that's a whole lot of hamburger lying there. I think we will go ahead and butcher her out and see if it is any good."

"Late Saturday afternoon, you're probably not going to be able to get a mobile slaughter out this afternoon."

"I will get the tractor out. It has the front-end loader attached. I can hang her here and have Daryl come pick her up in the morning."

"Okay, if you're going to do it, I will give you a hand. I sort of want to get a look at her back anyway."

Bart retrieved his tractor and his rifle.

Sandy and Marilyn busied themselves, getting ready for dinner and watching the kids, and we shot the cow.

It didn't take long before we had the carcass hanging from the tractor's elevated front-end loader. Bart was working on skinning the cow, and I examined the inside of the carcass.

There was an odd swelling on the underside of the backbone on the inside of the carcass.

"It looks like she must have had a significant injury to her back," I said as Bart looked over my shoulder. I point to the swelling with the knife I had in my hand.

Bart went back to work. He was almost done with getting the hide off the carcass. I reached up and ran my knife down the underside of the spine on the midline.

When my knife sliced through the swelling, it exploded, spraying my face and beard with thick off-white pus.

"The diagnosis is a spinal abscess, Bart," I said. "I don't think you want to eat this cow."

Bart stepped around from behind the carcass and stifled a laugh.

"You are quite a sight," Bart said. "I think you are going to have to borrow our shower."

"Look at the hole where that pus came from," I said. "She must have fractured a vertebra, and then it abscessed. I saw a dog once with a similar lesion, but I have never seen anything like that in the cow."

"I guess I will just leave her hanging here and call Daryl and see if he wants to pick her up in the morning. He can send her to the rendering company. That way, Marilyn won't have to mess with it."

It was a long walk to the house with a beard full of pus. At least Bart had a shower I could borrow. I offered Sandy a peck on the cheek, but she declined.

It was a good dinner and good conversation, so the entire evening was not a bust.

Old Three Toes

When I was growing up in Coos County, one rarely encountered a coyote, except on the high ridges. We didn't think much about it at the time. That was just the way it was.

I remember the first coyote I saw, on the top of Sugarloaf Mountain, on a cold morning Jeep ride with Uncle Robert. Twenty years later, coyotes had moved into the valleys. They were heard regularly and encountered with little effort if hunting them. They had become a significant problem to sheep ranchers. An occasional brave one would come close enough to the barnyard to snatch a chicken.

My Uncle Duke's explanation for the change was probably the most accurate. I didn't have a complete understanding at the time but would later come to appreciate his wisdom. In my younger years, all the creeks in the area were full of spawning salmon and steelhead in the fall and winter. Dead, spawned out, fish were present on the riverbanks and all the creek banks. Later in the 1950s and 1960s, commercial fishing for salmon moved from the streams to the ocean.

Spawning fish numbers decreased, and dead fish were only occasionally encountered on most streams.

Duke's opinion was that when the streams were chuck full of fish, the coyotes would have easy access to salmon and die from the disease. The only viable populations thus existed on the high ridges far removed from the spawning streams.

Salmon disease (or poisoning) is a complex disease of all canines. It occurs approximately seven days after a dog (or coyote) consumes infected raw salmon, trout, or steelhead. The fish carry a larva of an intestinal fluke. The fluke causes only mild disease and can infect several species, but the fluke also carries a rickettsia. It is this rickettsia that makes all canines ill and is the cause of salmon disease.

Salmon disease is treatable if it is caught in time. Over ninety percent of dogs (and coyotes) will die within seven to ten days of becoming sick if they are not treated. Survivors may be immune for long periods if not for a lifetime, although there are exceptions to this immunity.

***

I was on a farm call, talking with Dick Rice. Dick owned a ranch on the Calapooia River. His ranch was one of the early pioneer ranches in the area.

"Doc, I have been having a heck of a problem with coyotes the last couple of years," Dick said. "It seems to be the same coyote most of the time. He has only three toes on one foot. He catches any lamb left out of the barn overnight. Can't trap him. He is too wise."

Dick was at his wits end on how to deal with this bandit. I related my Uncle Duke's opinion on the shift of the coyote population into the western valleys. He listened with interest but just seemed to take it in as a story. I finished with the calf we were treating, loaded up, and returned to the clinic.

***

I never gave the conversation much thought after that until I bumped into Dick outside of Thriftway one afternoon. Dick had hurried to catch up to me in the parking lot. It was apparent that he wanted to talk.

"Hi Doc, how have you been?" he said, a little out of breath. "I have wanted to talk to you about that Old Three Toes.

"Aw, yes, I remember you talking about him," I replied.

"You know, I got thinking about the story you told about salmon poisoning. One night after work, I stopped in here and bought a hunk of a salmon tail. I have an old burn pit and garbage pile on the far side of the pasture behind the house. I took that salmon out there and put it on the edge of that pile. It was gone the next morning."

"And Doc, that was a couple of months ago. I have had no more coyote problems, and Old Three Toes is gone. I have not seen his tracks anywhere. Can't thank you enough for that story."

"I'm glad it helped you, Dick. You can thank the observation skills of an old farmer for the information. I am not sure that I would have ever put that information together to come up with that conclusion," I replied.

Too Many Legs

I pulled through the open gate to the pasture. It was early evening, the weather was great, one of those early spring days that we see in the Willamette Valley. Bright sunshine, no wind, and pleasant moderate temperatures, probably in the high sixties. In past years, this was the type of day that I would be skipping school to go fishing.

I could remember Mel's words on the phone. I was just hoping he was correct.

"Doc, this is Mel, out on Pleasant Valley," Mel said into the phone. "I have a heifer down out in the pasture. She has been down and straining for a couple of hours. I don't think she can get up. I left the gate open for you. I have to go to work. I would appreciate it if you could take care of her. Just leave a note in the mailbox when you are done. I will call you in the morning."

These heifers often became wild again when a strange pickup showed up with a stranger driving. Just to be safe, I got out and closed the gate. I could still hear my grandfather.

"It is a lot easier to close the gate than it is to wish you had closed it," he often said.

I pulled up to the heifer. She made no effort to move. I could see front feet and the calf's nose sticking out of her vulva. It looked like a normal position. It must be a large calf.

I got out of the truck and poured a bucket of warm water. I put a rope around her neck. There was nothing to tie it to except my truck. I had learned that lesson a few years earlier. When a cow is tied to the truck, she goes in a circle on the end of the rope. This usually means she collides with the side of the truck somewhere. That typically leaves a big dent.

Today I tie the rope to her front leg, bringing the foot up close to her neck. I think that should keep her from getting up long enough for me to get control of her. After I have her restrained, I tie her tail out of the way with some twine. And then prep her rear end.

The heifer is a Hereford and the calf looks like a black baldy. She is black, with a white face. Usually, these calves are crosses between an Angus and a Hereford. They generally grow up to be good cows. The crossbreeding provides some hybrid vigor.

This heifer is young, less than 2 years old. She looks like she should be large enough to deliver this calf.

I put on an OB sleeve and applied a lot of J-Lube. I ran my hand into the vagina alongside the calf, first on the right side and then on the left side. Everything felt fine to me. There appeared to be plenty of room in the pelvis, and the calf was in a normal position. The calf was still alive.

I thought we just needed a little traction on the calf, and it should just pop out of there. I put my Frank's Calf Puller together and seated the breach across her hind legs below the vulva. I hooked the feet to the puller with a nylon OB strap. Then I started jacking the calf out.

There was minimal progress, and then it came to a solid stop. I applied a little more pressure, nothing. These calf pullers were sort of two-edged swords. They did the job easier, but they also allowed someone to put too much force on the calf. This was dangerous to the survival of the calf. It was also hazardous to the tissues and the nerves of the momma cow. The idea was to put no more pressure than two good men could apply. That was sort of a learned skill.

I stopped and unhooked the calf and set the puller to the side. Then I gave a quick wash to the vulva again. I explored the birth canal again, bare-armed this time, so I don't lose any sensitivity due to the plastic sleeve on my fingers. I could not feel anything that would be a problem with this delivery.

I hooked up the calf puller again and put tension on the calf. Then, I pulled the end of the puller down, so it was putting a downward pull on the calf. This would also make the breach put upward pressure on the calf. This did a couple of things. It gave the calf a direction of travel as if the cow was standing. This also elevated the calf in the birth canal. The pelvis was a little wider in the upper portion of the birth canal.

I was thinking that I was putting too much pressure on this calf. But I gave a little more pull down on the end of the puller. It was close to vertical relative to a standing cow. I was about to stop when I detected a slight slippage of the calf. I gave one more little pull on the end of the puller.

The calf suddenly came out, almost as if it was shot from a cannon. Landing on the ground, the calf shook his head. At least he was still alive. The heifer groaned a little, but I am sure she was relieved that the calf was out of there.

I went over to look at the calf. He was holding his head up already. And then I saw the problem. This calf had an extra set of legs coming out of his back just behind the shoulder blades. These came up out of the back and folded back along the back. They just added enough extra depth to the calf's chest to make it a tight fit for the birth canal.

Doing a quick exam of the calf, his hind legs were paralyzed, a lot of effort for the momma cow, all to no avail. Now, what to do. There is nobody home. Mel is at work. I don't know if I have a phone number to reach him. In any case, it will involve a trip back to the office to make the call.

It would be interesting to know just what is going on structurally with these extra legs. I might see if Mel has any interest tomorrow. I am sure Mel would concur that this calf has no future. Leaving him until tomorrow will just add more stress for both the calf and the momma cow. I take a deep breath and decide to go ahead and put the calf to sleep now.

I draw up ten ccs of Sleep Away and give it IV to the calf. He is gone before the injection is complete. I return to the heifer and check her birth canal for any injury. It seems fine. I give the membranes a little tug, and they come out with little effort. I instill 5 grams of tetracycline powder into her uterus because of the extended labor.

Now to see if she can get up. I untie the rope from her foot and take it off her head. I stand back and give her a swat on the rear. She jumps up with no problem. She only glances at the calf before she heads off to the far end of the field.

I pull off my coveralls and pour a fresh bucket of water to wash up. Sandy wanted me to stop at the store for a couple of items. I was going to have to hurry if I was going to make it to dinner.

I left a note for Mel, saying that I will talk to him when he calls in the morning and letting him know the calf is dead in the field. In the note, I tell him real briefly that the calf had six legs. That was why the heifer needed some help. I also say to him that the calf was paralyzed and that I put it to sleep.

I stop at Thriftway for a loaf of bread and a gallon of milk. I much prefer Thriftway for their service, but also their community support. It is great to have a locally owned, large grocery store in town.

I find it a little odd that people are avoiding me in the store. Maybe it is because I am almost running to get things and get checked out as soon as possible. But I make it home just as Sandy is getting the kids sat down for dinner.

Sandy gasps as I am putting the milk into the refrigerator.

"Did you go to the store like that?" she asked.

"Like what?" I say.

"You go look in the mirror in the bathroom, and you wash before you come to the table!" Sandy says in a firm voice.

I look myself over in the mirror, and I don't see anything that I would consider unusual. I know there have been times when I have missed blood in my hair and the like, but today I don't see anything.

Sandy comes up behind and touches the back of my elbow. I raise my arm and look at the back of my elbow. There is a large swab of thick mucus and blood covering the back of my arm and elbow. No wonder I was avoided in the store.

Cookie's Litter

"Dr. Larsen, this is Maude, from Brownsville," Maude said into the phone. "Cookie, my best milk goat, is about ready to deliver. She is so large that she has been down for two days. Can I bring her up to your place so you can look at her?"

"That's fine, Maude. You just need to know that we are not in the clinic yet," I said. "We are still practicing out of our house on Ames Creek."

"I think I can find it," Maude said. "It doesn't bother me, and Cookie is used to an old barn, so a garage won't be a problem for her."

When Maude arrived, and we got Cookie unloaded, her appearance was amazing. She had been down for two days because her abdomen was so massive, she could not support it standing. I was worried that she might even have hydrops allantois, but I could easily palpate a couple of very active kids in the uterus.

The only time I had seen hydrops was while I was in school at Colorado State University, in a cow bred to a bison bull. In that cow, you could not palpate any fetus. And on C-section, the calf was dead. Hydrops is common in cows bred to bison bulls. The beefalo calves come from bison cows bred to a bovine bull.

"We have some decisions to make, Maude," I said. "If we wait for her to deliver, there is a good possibility that she will suffer significant musculoskeletal injury from being down for such a long time. If we do a C-section, we take a chance on the kids being early. That will mean that not all of them may survive."

"You say, all of them," Maude said. "How many do you think are in there?"

"I think a bunch," I said. "I have seen four lambs in a ewe, and she was almost this big. That is about a one in ten thousand chance, and it might even be higher in the goat. There are at least three kids in there. I think there may be four."

"I think I am more worried about Cookie than I am about the kids," Maude said. "It would be nice to get a bunch of kids out of her, but I want you to know that Cookie is the priority in this event."

"So, I hear you say that you want to do a C-section," I said.

"Yes, that is what I want to do," Maude said. "When can you do it?"

"Maude, I am just getting started here," I said. "My days are not full. I can do it right now."

"Good, can I stay and watch?" Maude asked.

"You are more than welcome to watch," I said. "In fact, you might be put to work if we have four kids in there. I see you have a couple of bales of straw in the back of your pickup. Is there a chance we could use one of those to bed her down? She would be more comfortable than on the bare concrete."

"That is what I brought them for," Maude said. "Let me pull them out of the truck."

We used one of the straw bales to bed Cookie down in the back of the garage while we set up for surgery. Sandy set some chairs out for Maude and her driver.

"Once we get set up, things will go pretty fast," I explained to Maude. "I will roll her up on her back, and we will clip and prep an area in front of her udder. I will make an incision, then we will start pulling kids out as fast as I can. If there are four of them, everybody will have a kid to take care of as I close up things for Cookie."

"I have never seen anything like this before," Maude said. "I guess I will be okay. At least if there are kids to care for, I will have something to do."

Paula had everything ready, so we rolled Cookie up on her back. Again, I was amazed at the size of her belly. It spread out in both directions.

"I don't think she could roll off her back if she tried," I said to Paula. "You better tell Sandy to get every spare towel she can find. We are going to spill all sorts of fluid out of this uterus."

With the abdomen clipped and prepped, I made the incision on the ventral midline in front of the udder. My kids watched from the kitchen doorway, and Maude sort of stretched her neck to see better.

"I thought there would be more blood," Maude said.

"As long as I can avoid these large milk veins coming from the udder, there should be very little blood," I said.

I extended the incision through the linea alba, pulled the omentum forward so it was out of the way. Then I reached in to grab the head of the first kid I encountered. I drew this head through the abdomen incision and incised the uterus carefully.

The kid's head popped out of the incision and shook. I think she was ready to be out of there. Grabbing her neck, I pulled her the rest of the way through the incision. There was a rush of fluid that came with her. I handed her back to Maude. Maude was waiting with a towel.

"This kid is the same size as a single," Maude said.

"No wonder she is as big as a house," I said as I reached in and grabbed the second kid by the head also.

This kid came out fighting also. And like her sister, she was the size of a single.

"Two girls so far," I said. "In a couple of years, you're going to have so much milk you won't know what to do with it."

I reached in and grabbed a couple of hind legs in the far uterine horn. I tugged, and they did not move a lot. I felt close. I had one leg from two different kids. Correcting my error, I pulled the third girl out by her hind legs.

"Maude, there is another one in there," I said. "And this one is another girl."

Sandy stepped up, with the help of our girls, to take this kid. Maude had the first kid standing already. Everybody was busy now, and there were smiles all around. Nothing like baby goats to make people happy.

I reached in and pulled out the last kid, another girl.

"This probably sets some sort of a record. Quadruplets, and all females," I said as I handed the last kid to Paula.

I started pulling as much of the fetal membranes out through the incision as I could. There is no way to take them all out without damaging the uterus. But I would pull out a bunch and cut them off with scissors. We had all the towels and straw soaked with uterine fluids.

"This is probably going to take as long to clean up this mess as it took for the surgery," I said, more to myself than to anybody else.

I closed the single incision in the uterus. It was about a six-inch incision. It was sort of amazing that you could drag four kids through that incision. After returning the uterus to a normal position, I closed the midline with a sliding mattress suture using number two Dexon. The external incision was closed in a conventional two-layer manner.

"Cookie, you are going to have a flabby belly for a time," I said as we rolled her off her back and made her comfortable on a clean spot in the straw. She wanted the kids, so Maude started handing them over to her one at a time. It didn't seem to bother her at all to have all four of them to care for.

"What do you think we should do with her now, Doc?" Maude asked.

"I think we should milk her out and get some colostrum into the kids," I said. "It is early enough that leaving the kids with her for the day will probably be the best. I will give her some intravenous glucose and calcium just to give her a little more energy. I am guessing that she isn't going to be up until morning. I would like to keep her until then. I don't know if I can handle the four kids running around the house."

"I will go home for the day," Maude said. "I have a lot of chores waiting. I will get back here around five this afternoon if that's okay. I will take the kids home. That is not a big thing. I have plenty of colostrum in the freezer, and I don't leave the kids with my milkers anyway."

"That will work fine," I said. "Then, we will just see what morning gives us with Cookie."

***

Cookie was up and looking for both food and a milker when I checked her early the following day. Luckily, Maude pulled into the driveway about when I was heading into the house to give her a call.

"I guessed that Cookie would be up and ready to be milked," Maude said. "I figured you would appreciate getting her out of here early."

"I think she will be ready for the milking stand by the time you get home with her," I said. "She should be good to go. I see her membranes passed last night. Just keep an eye on her, and I will drop by to just glance at that incision in a couple of days. I generally leave those sutures in place for about three weeks, just to be safe."

Cookie healed with no problems, and the kids became a great addition to Maude's milking string.

Long Road Home for Tramp

"Slow down a minute, Ralph," Jan said as she was watching the old cat on the edge of Pleasant Valley bridge in Sweet Home. "Turn here," Jan said, pointing at the bridge.

Ralph turned and headed across the old bridge.

"Stop, stop right here. That cat needs some help."

Jan almost jumped out of the car before it came to a stop. She crouched down and called softly to the cat. "Here, kitty, kitty," Jan said as she stretched out her hand and made a couple of short, shuffled steps toward the dusty old tabby cat.

The cat hesitated for a moment as if trying to decide if he should run or not. But something was inviting in this lady's voice. He eased forward and sniffed at her fingertips. She patted him on the top of his head.

A couple of cars had stopped behind their vehicle, and Ralph was getting a little impatient.

"Hurry it up, Jan. We are holding up traffic."

A lady started to get out of a car that was a couple of cars back in the line. Jan motioned for her to stay back.

49

The old tabby cat moved up to Jan's knees and pushed against her.

Jan could feel a stifled purr. She took a deep breath, leaned over, and scooped the old guy up.

Jan slid into the car with the cat and pulled the door shut. The cat leaned into her and purred as Jan stroked his back and sides.

Ralph swallowed and put the car in gear. "I hope this isn't a mistake," he said as the car moved forward.

"This is a nice cat," Jan said. "And he has a collar and a tag."

"We don't have time to deal with a stray cat today," Ralph said.

"We need to find the vet's office in town," Jan said.

Ralph pulled over as soon as they were across the bridge. The car with the lady who wanted to help pulled up behind them, and the lady came up to Jan's window. The cat was now wholly under Jan's spell as she continued to stroke him with long slow strokes from the top of his head to his tail.

Jan rolled her window down a bit. "Where can we find a vet in town?" she asked.

"There is a clinic in the Safeway shopping center in the middle of town," the lady said. "Is the kitty okay?"

"I think he is okay, maybe lost, but okay," Jan said. "He looks a little rough like he has been traveling a bit. He has a tag. We will drop him at the vet's office. We are headed for Bend and don't have a lot of time."

***

Jan was breathless as she came through the clinic door and perched the cat on the counter in front of Judy.

"We found this cat on the bridge coming into town," Jan said. "It looks like he needs some help, and we are on our way to Bend."

"It looks like he has a tag on that collar," Judy said. "Is he nice?"

"He is the sweetest old thing," Jan said. "I think he must be lost."

Judy looked at the tag. "It says Tramp," Judy read. "I guess that fits. Let me check with the doctor."

I came out and looked at the cat. He was thin but okay otherwise, and it had a collar and a tag. The tag gave the cat's name, Tramp. It also had an owner's name and local phone number. I agreed to keep the cat.

"Thanks a lot, Doc," Ralph said. "We have to hurry now. We have a meeting in Bend that we will be late for if we don't get on the road."

These foundlings were always a problem. Occasionally, the finder would offer to be responsible for the bill if the owner was not found. But most of the time, that expense, whatever it happened to be, fell on the clinic. At least Tramp came with an owner's name and phone number.

Judy was given the task of calling the owner on the tag.

"Yes, this is Robert Wilson," the man said to Judy. "What can I do for you."

"This is Judy from Sweet Home Veterinary Clinic," Judy said. "We had a couple find an old cat on Pleasant Valley bridge this morning. The cat has a tag on its collar with your name and number on the tag."

"I don't know what to tell you about that," Mr. Wilson said. "We don't own a cat."

That was great news. We were stuck with finding someone to adopt this cat, not an unusual event for such situations.

About thirty minutes later, we were still discussing how we would find someone to take the cat, and the phone rang. It was Mr. Wilson, the guy Judy had called about the cat.

"What does that cat look like?" He asked.

"It is a brownish tabby cat, neutered male, friendly. He looks a little thin and has sort of a rough hair coat, but otherwise, he is in good shape." Judy replied.

"We had a cat about five years ago. We had to move to San Francisco for a couple of years. We lost him on the trip down, somewhere in northern California. His name was Tramp, but I don't remember a collar. You don't think that could be him, do you?"

"How else do you suppose this cat had Tramp's collar?" Judy asked.

"We will come right down and get a look at him."

It was not long, and a car pulled up in front of the clinic. Robert and his wife came through the door first, but Susie, their teenage daughter, was right on their heels.

One look at Tramp, and it became a happy reunion. The daughter opened the kennel, and Tramp was instantly on her shoulder and purring, rubbing his face on her neck and face. She was in tears.

"Susie has suffered for years. On the trip to San Francisco we had stopped at a rest stop south of Crescent City, and Tramp got out of the car. The next thing we knew, he was scared by another car and ran into the woods. We looked for him for an hour, but we couldn't stay there. We had to go on. Susie cried for days."

"Do you think he has been traveling all these years? That is remarkable," Judy said.

"It is pretty hard to believe, but you saw the immediate recognition by both of them. Pretty remarkable, but it will be a happy evening in our house," Mr. Wilson said. "Do I owe you guys anything?"

"No, we are just happy we didn't have to find a family to adopt him," I said.

The stories Tramp could tell. This was something right out of a Disney movie.

One More Pregnancy

"I wish Dr. Ball was here," I thought to myself as I drove my left arm deeper into the rectum of this old fat cow. Dr. Ball was the chief OB instructor at Colorado State when I was in school there. He was tall and thin, and it seemed that his arms were long enough to reach his knees, and his fingers were at least six inches long. He could do anything in a rectal exam.

"How am I going to get this uterus retracted," I thought as I swept my hand across the floor of her pelvis. I could just grasp the cervix; the rest of the uterus hung over the brim of the pelvis and was heavy and resisted my attempts to pull it back so I could get a better grip.

"Bob, I don't know if I am going to be able to retract this uterus," I said as I looked at an anxious owner. "Let's review her history one more time."

I pulled my arm out and removed the OB sleeve. I could do a hundred pregnancy exams in a couple of hours, but this old cow had already tired my arm out. I leaned on the railing of the crowding alley, resting my arm, as I talked with Bob.

"How old is this cow?" I asked.

"She is going to be nineteen," Bob answered. "She is my first cow and she is the genetic base of my entire herd. I realize it is a big request, but I would love to have another calf out of her."

"When was the last time she had a calf?" I asked.

"Three years ago, probably a little over that," Bob said.

"When was the last time you bred her?" I asked. "I mean, is there a possibility that she has a pregnancy. I have not even been able to get the uterus up where I can check for a membrane slip."

"No, I quit wasting semen on her well over a year ago," Bob said. "I didn't have a bull before we moved. This young guy I have now might get a chance at her if you say go."

"So, this is the situation," I started to explain. "This cow is nineteen, very over-conditioned, fat in other words. She hasn't been pregnant in over three years and has a uterus that is out of reach. That most likely means there is some chronic pathology in that uterus. That all adds up to a slim chance of getting her pregnant."

"What is the first step in trying?" Bob asked.

"The first step is to retract her uterus so we can decide what might be going on inside of it," I said, a little unsure that I had been clear enough in my explanation. "In school, we had Dr. Ball. He could retract this uterus in a heartbeat. Without Dr. Ball, we used a large cervical forceps that could clamp on the cervix to help pull the uterus up where you could get hold of it. I don't have such a forceps. But I have a large Oschner forceps that has some teeth on it and might be able to provide a little traction. The only problem would be it could put a small tear on the cervix if it doesn't work."

"If that is what it takes, let's do it," Bob said. "This is probably the last time we have a chance of getting her pregnant."

"Just to be clear, that chance is less than twenty-five percent," I said. Why did I give such a figure? Never give odds to a horse owner, or probably a purebred cow breeder. They will take you up on it, every time.

I tied the cow's tail out of the way, tying it with a loop of twine around her neck. That way, if we forgot to untie it before releasing her from the chute, we would not pull the end of her tail off.

After prepping her vulva, I tied a length of gauze to the handle of my twelve-inch Oschner forceps and carried the forceps into the vagina with my left hand and arm. Along with the forceps, I brought a guarded culture swab.

The cervix was large but felt relatively healthy and was tightly closed. I advanced the guarded swab to the cervical opening and pushed the swab into the uterus. I pulled the swab back into the guard, pulled that out with my right hand, and handed it to Bob.

"Hold onto this for a minute," I said.

Then, back inside, I clamped the forceps on a stout external ring of the cervix at the top. Then I tested the gauze with a firm pull. The forceps held, and the cervix did retract a little. I withdrew my hand and made sure the gauze followed and hung outside the vulva.

"Bob, do you know if this cow cycles normally?" I asked.

"Yes, she cycles as regular as can be," Bob said. "You can almost set your calendar on her every twenty-one days."

"Do you know when she is due to cycle?" I asked.

"She should cycle in the next few days," Bob said.

I changed the sleeve on my left arm. I pulled the fingers off an OB sleeve, pulled it on my arm, and then stretched on a latex exam glove, covering it again with another OB sleeve with the fingers removed. This would provide much better sensitivity in my fingertips.

"This is the plan," I started. "If this uterus feels like it is salvageable at all, I will infuse it today with an antibiotic. Then if she cycles, you go ahead and breed her with the bull and call me. We will infuse her again, twenty-four hours after breeding."

"I would rather breed her with artificial insemination semen and save the bull for the last resort," Bob said.

"Okay, I just want to try on this cycle, just in case we get lucky," I said. "We will have the culture results back by next week and be able to make an antibiotic selection based on that culture. If she doesn't get pregnant, we will go through a series of infusions. We might be working on her for several months."

I drew an infusion of Furcin solution and neomycin into a 60 cc syringe. I attached an infusion pipette to the syringe. I handed this to Bob, and I took the culture swab and set it aside. That done, I lubed my left hand and arm and pushed as deep into the rectum as I could. I grabbed the cervix and, at the same time, with my right hand, pulled on the gauze attached to the forceps. The cervix and uterine body retracted into the pelvis. I reached forward and flipped the entire uterus back into the pelvis — sort of amazed myself.

"This uterus is the size of a uterus with a ninety-day pregnancy," I said.

Out of habit, I checked for a membrane slip; there was no pregnancy. The uterus was thick and dough-like, with no significant fluid present. Probably just a case of very chronic uterine infection, endometritis, I thought. The ovaries were active and felt normal, with a receding corpus luteum on one ovary.

"Okay, hand me that syringe," I said.

Holding the cervix, I worked the pipette tip through the cervical rings and made the infusion. Then I allowed the uterus to return to its position. I pulled my arm out and removed the sleeve, carefully turning it inside out. I prepped the vulva again. Then with a new sleeve, I retrieved the forceps from the vagina.

And so it began, a months-long battle with an old, chronically infected uterus, trying to bring about a pregnancy. I went through a week of antibiotic infusions based on culture results. We used post-breeding infusions after every breeding attempt. We used artificial insemination, and we bred her with the bull. She continued to cycle regularly. At least she didn't raise any hopes by missing a heat period.

I was at wit's end, and then I remembered a conversation with Don in Enumclaw, on our way back to the clinic, from doing some pregnancy exams. Don enjoyed picking my mind. We had gone to different schools, and sometimes our views on how to do things were completely different. We both modified our thinking in small ways from our discussions.

"I want to tell you," Don started, "if you ever get in a situation where you want to get a cow pregnant, I mean, if you have tried everything and you really want her pregnant, you infuse her with dilute Logul's.

"That is probably the only absolute 'never do' that Dr. Ball told us," I said. "That is supposed to do far more harm than good."

"I am just saying," Don repeated, "if you really want to get that cow pregnant, use an infusion of dilute Logul's. It works, maybe not every time, but it works."

"I don't know," I replied, somewhat skeptical of the advice.

"You do an infusion; the cow will cycle in six - eight days, skip that cycle and breed her on the next cycle. You can do a post-breeding infusion if you like. You will have a seventy or eighty percent conception rate on that breeding. And most importantly, you will have a pleased client. I know what the literature says. And I know how you were taught, but in this case, my experience trumps all of that. Just file it away in your mind and try it if the need arises."

On my next trip to check the cow, the uterus was improved. After all the treatment, it sort of felt like a normal uterus now. I could retract it into the pelvis in a pretty standard manner now. So things were improved, but still, no pregnancy resulted.

"Bob, I have one more trick we could try," I started as I explained the new plan. "There is a treatment that I was taught in school never to use, but the guy I practiced with when I was in Enumclaw swore by the results in just this situation. We have pretty much exhausted the book. We have worked through everything, and the uterus is better. You can probably remember the difficulty I had during her first exam. But today, the exam goes easy.

"What I suggest is an old-time treatment. I do an infusion with dilute Logul's. Logul's is an old time solution of iodine and potassium iodide. This infusion wipes out the lining of the uterus. And then the lining of the uterus is regenerated from a few remaining cells."

"Sort of like a chemical D&C," Bob said.

"Yes, I guess that is a good way to look at it," I said.

"It doesn't look like we have any other choice," Bob said. "Let's do it."

It took me almost a week to get a bottle of Logul's. It was not something in the regular supply line. When it came, I had no recipe for mixing a dilute solution. I sort of went by the old port wine formula that I used for Betadine.

"It should look like port wine" Dr. Annes had always said.

The funny thing, I doubt if any of us students had ever seen a glass of port wine.

I mixed a solution, and we gave Bob a call and set up a visit.

The infusion was no problem. The old cow was used to the chute now. I noticed that when we turned her out of the chute and started discussing the breeding schedule, she was hunched up a little. There must have been some significant discomfort with the infusion.

"Expect her to cycle in eight days or less," I said. "Don't breed her on this cycle but breed her on the following cycle. We should plan a post-breeding infusion just to be on the safe side."

***

Her cycles went right by the book, eight days and twenty-one days after that. Bob called when he had bred her. I had hoped he would use the bull, but he used artificial insemination. He probably figured he had gone through all this trouble; he wanted a good calf.

I marked my calendar for her next expected cycle. The date came and went, and there was no call from Bob. Too early to celebrate, I thought, but just maybe, she is pregnant. At forty-two days after breeding, Bob called.

"She still hasn't cycled," Bob said with some excitement in his voice. "When do you want to check her?"

We waited until fifty days. With her uterus being larger than most, I didn't want to be in a situation where there was a question with the exam.

When 50 days came, the exam was brief. The uterus felt the best that I had seen in this old girl, and a 50-day pregnancy was present based on a membrane slip and palpation of the amnion.

I was relieved and happy. Bob was happy. And I am sure the old cow would enjoy not being run through the chute every few days. And the bottle of Lugol's probably sat on my shelf for the next twenty years before I discarded it.

Polyradiculoneuritis (Coonhound Paralysis)

Neurology in veterinary medicine in the early nineteen seventies always seemed like a waste of time to me. In school, we seemed to spend untold hours learning detailed anatomy of the brain and nervous system. There were some things we could intervene in, like spinal injuries. But the viral diseases of the brain and even the bacterial infections were tough nuts to crack in veterinary medicine in 1970.

It seems every neurologist loved their specialty. Somewhere during every series of lectures, polyradiculoneuritis would pop up.

After spending the better part of an hour on the topic, the professor would note that most of us would never see a case in our lifetime. Why then do I have pages of notes on a disease that I will probably never see? If I do, there will be virtually nothing that I can do for the patient except provide competent nursing care and hope for a recovery. Of course, I had the notes because of the pending test. I always felt that it was a total waste of time.

In practice, I ventured into the brain only on rare occasions. My first case was a necropsy on a cow who died suddenly during the morning milking. I was a budding pathologist in those days. I completed a very thorough necropsy, only to find absolutely no reason for this cow to die. The owner wanted an answer, and the only place I hadn't looked was the brain.

I had spent the summer following my sophomore year in vet school working on the necropsy floor at school. Extracting the brain was an easy task for me. I skinned the head and then shaved the bone away from the brain with a small cleaver. A couple of snips at the dura and I lifted the brain out of the skull. Laying it on a board, I sliced it in thick longitudinal layers. On the third slice, I hit a large pocket of mush. This cow had a massive stroke. I offered to send in the tissues for an accurate diagnosis, but the farmer was aware of our limitations.

"What are you going to do with any answer they give you?" he asked.

***

Then there was Buddy. Buddy was a twelve-week-old hound pup belonging to Frank Updegrave. I had seen Buddy a couple of times for routine vaccines and such. On this day, Frank was helping a friend build a shed. They were putting up rafters. Buddy was running around the shed doing hound stuff. Like any good hound, his nose was to the ground as he followed some scent. About then, he was suddenly in the wrong place at the wrong time. One of the rafters fell from the top of the roof. The end landed on the front of Buddy's forehead. They gathered him up and came running to the clinic.

The top of Buddy's skull was caved in, depressed into the brain, maybe an inch. The frontal sinuses were open. And worst of all, I could see brain tissue oozing into the wound.

After just a brief look, I took a deep breath and turned to Frank with an assessment.

"Frank, I don't think I am going to be able to help him. His skull is caved into his brain, and the frontal sinuses are open to the wound. The chances of saving this guy are slim, in fact, slim to none." I said.

"I know, but we have to try, Doc. Can you just try? I have total faith in your skills," Frank replied.

In those days, we didn't have specialty clinics on every corner. If this were going to get done, it would be by my hands.

"I will give it a try. I will do everything I can. Just one thing, Frank, I want you to sign a euthanasia release before surgery. If things go from bad to worse, there is no reason to put Buddy through any discomfort by waking him up."

Buddy was unconscious through the entire exam. I intubated him without induction drugs and used only halothane gas anesthesia.

We shaved and prepped the wound and reflected the skin edges from the bone. There was probably a two-inch square of skull bone depressed nearly an inch into the brain. Brain tissue was oozing around the edges of the depressed bone. The break of the skull bone opened the upper corner of the frontal sinuses.

I was obviously beyond my experience base at this point. I used curved mosquito forceps to pry the depressed bone from the brain. I mopped up the loose brain tissue and wondered what to do next. There was a significant depression in the frontal lobes of the brain. I placed a couple of sutures in the dura mater, just enough to close the tear. Then I placed a couple of twenty-two gauge stainless steel sutures in the leading edge of the skull bone to maintain a substantial reduction. The posterior portion of this bone was still attached. It was a jagged enough leading edge to provide a reduction with adequate closure of the sinuses. I routinely closed the skin wound. Then I unhooked Buddy from the gas. Now it was just a waiting game.

By the end of the day of surgery, Buddy was becoming responsive. He would acknowledge your presence, even raise his head a little. The next day, Buddy was sternal but pressing his right side to the edge of the kennel. On the second day, he would eat a few bites and walk as long as he had a wall to press against his right side. That meant that he could move but only in a counterclockwise direction around the room. Another couple of days, and we sent Buddy home. He could walk now without a wall and improving every day.

The following summer Buddy was scheduled for his annual exam. I glanced out into the waiting room. Frank, a tall, lanky young man who may be a couple of years out of high school, was sitting with Buddy on his lap. Buddy was a full-grown hound and took up his entire lap. Buddy was a forever puppy. Pretty functional, but he never progressed beyond his ten or twelve-week-old mental abilities.

***

It was maybe a couple of years after Buddy's accident that Sally Smith brought her Australian shepherd named Sport in for an exam. Sally was a small blonde lady in her early forties, very athletic. She had purchased the BeBee brothers ranch in Liberty and ran emus Sport was a cow dog who had only birds to herd.

During the exam, Sport displayed a lot of anxiety. It was apparent he knew that things were not right. All of his reflexes were impaired. He could barely walk. Sally said he seemed to worsen by the hour. My first concern was rabies, but his vaccination was current. Sally agreed to leave him overnight for observation and treatment.

I was at a loss for a diagnosis. I started Sport on some trimethoprim/sulfa and some Dexamethasone to cover bases. The trimeth/sulfa would treat toxoplasmosis specifically and would also cover most bacterial infections of the central nervous system. The steroid would reduce and central nervous system swelling. We would have to see what morning would bring.

By morning Sport was completely paralyzed. He was flat out, could not raise his head, but was bright and alert, following my every move with his eyes. He could lap water and eat with assistance. Without a diagnosis, I was dead in the water to provide a prognosis, but things were looking pretty bleak at this point.

Sally came in later in the morning, and I accompanied her back to the kennel to look at Sport. Sport laid there with his eyes dancing and his mouth open and tongue lapping. The tip of his tail still had a slight wag. I knelt and patted his head. Then I noticed a large scratch across the top of his head.

"That's quite a scratch," I said.

"Oh yes, that is probably from a raccoon. Buddy is out hunting those things every night," Sally replied.

I stood up and looked at Sally, "Raccoons," I said.

"He runs them all night long and often tangles with them. He killed a big old boar the other night," she said.

I knelt and scratched Sport on the head, thinking to myself, "Wow, polyradiculoneuritis."

Don't you believe it! I said to myself, "I'll be damned, Buddy, you have coonhound paralysis!"

Apparently, I spoke louder than to myself. Sally asked, "What did you say?"

"This is coonhound paralysis. It is a rare disease in dogs, very similar to Guillain-Barre syndrome in man. We don't know what causes it, but it is most often associated with contact with raccoon saliva, hence coonhound paralysis.

"There is not much to do for him except to provide nursing care. It will get worse; all his muscles will atrophy. But if his respiratory muscles remain functional, there is a chance that he will recover and return to normal. We can care for him here if you like, but the expense may be high."

"I will take him home and make a bed for him behind the stove. He will be much happier at home," Sally said.

"Make sure his bed is well padded, turn him often, several times a day, and help him with food and water. We will try to keep track of things with you but call if you have any questions," I replied.

It was several months before I saw Sport again. He was pretty much back to normal. Sally said he was looking pretty bad after a couple of weeks of paralysis but then slowly returned to normal. Now he is back to chasing those darn raccoons.

***

About two years later, Sport was in the clinic with another onset of paralysis. Things went about the same as his initial episode. The neurologists in school had always said it would be a once-in-a-lifetime diagnosis. I would guess that I must have lived a couple of lifetimes. The diagnosis was made twice but to the same patient.

A Perfect Delivery

I glanced out in the waiting area and could see Emma waiting to talk with me. Emma was an attractive young girl with light brown shoulder-length hair that she wore in a ponytail. I think she was still in high school, probably a junior or senior.

Emma had a young mare, Lilly, due to foal most any time now. Emma was doing everything in her power to provide the perfect setting for the delivery. In doing so, she has been talking my leg off. She had been talking with me a couple of times a week for the last month. Most of the time, that was okay. I did a lot of work for her father.

When I stepped out to the front counter, she bounced up.

"My father says I have been bothering you too much and not paying you a fair fee," she said. "So I want you to make a farm call and check out the birthing facility I have set up. I have moved a bed into the barn, and I will be sleeping there until Lilly foals."

"You tell your father I am always willing to provide whatever instruction I am capable of to our clients and their families," I said.

"I know that, and so does he. We just thought maybe you should check over what I have set up to make sure I am not missing something. I want this to be perfect."

"That sounds fine. You can schedule a time with Sandy. It is probably a good thing. We have covered a lot of topics over the last month or so."

"I have a checklist from those discussions and from reading in Horse and Rider magazines."

"You know, Emma, a lot of mares will be reluctant to foal if they are being watched," I said.

"But she is so ready. All the signs are there," Emma said. "She is leaking milk, and her privates are really swollen and flabby. And her due date is tomorrow. I am taking off school tomorrow, and Thursday, and Friday if she hasn't foaled by then."

"They have their own clock, and don't be surprised if you don't go to the house for dinner or something and come back to a foal standing in the stall," I said. "But, you schedule a time, and I can get out there this afternoon and see what you have set up."

The Pedersen farm was anything but neat. The barn was a large old barn, once painted red, set a hundred yards behind the house. With all the work on the farm, Mr. Pedersen didn't have a lot of extra time to worry about mowing the lawn. Emma was the oldest of five girls, and I don't think any of them helped around the barn much unless it was with Emma's horse.

I drove past the house and parked the truck by the barn. Emma came out of a small attached shed on the house side of the barn. Her younger sister was by her side, Sara was seven years old, and she was often around when we were working with the cows. Both girls were all smiles, and you could tell that the pending birth would be an exciting event for them.

I was literally blown away when Emma and Sara led me into the shed with the horse stall. It was immaculate. There was not a cobweb in the tallest rafter. She had a well-made cot in the corner with a desk and bookcase nearby. Then she had a small refrigerator on a shelf for medication and supplies.

Lilly was in a sizable stall that was bedded entirely with straw. There was a pitchfork by the stall gate and not a trace of soiled straw in the stall.

"Do you think the straw is clean enough?" Emma asked. "I have worried about that, but I don't know what else there is that I could use."

"The straw is fine," I said. "It is far better than most foals get."

"Emma thinks that it is going to be born tonight," Sara said. "I want to bring a sleeping bag out here, but Mom won't let me."

"Your mother is probably right," I said, "it is a school night. When mares have their babies, it is usually a pretty fast event. You would probably sleep right through it."

"I just worry about all the little things," Emma said. "The magazines talk about all sorts of problems. Things like navel infections I can feel confident that I can control by dipping the navel into iodine. They talk about foals suffocating in their membranes. Stuff like that where you have to be there to help, or you lose a foal.

"You have things just about as perfect as they can be, Emma," I said. "Those stories like the foal suffocating in the membranes are just stories. Most of those foals were probably stillbirths. Things happen fast when mares foal and most of the foals will not allow any membranes to hang around on their heads. Horses have been doing this a long time before people got involved in the process. Being here to watch is okay, but you don't want to do anything unless there is a problem. And then you should call me first if you can."

"Okay, I will relax a little," Emma said. "At least you have made me feel a little less concerned. It is just that I want everything to be perfect with this delivery."

"And Emma, don't worry if she doesn't foal tonight," I said. "Mares will often hold off their labor if there is too much observation. The big horse ranches usually monitor their mares in labor with remote cameras."

"Okay, but you know I am going to call you if anything looks unusual."

With that, I returned to the office and Emma sort of faded into the background for a time. Wednesday came and went with no call.

By Friday afternoon, I had just about forgotten about Emma and her mare. Then the phone rang.

Sandy answered the phone and quickly handed it to me. There was a very frantic Emma on the other end of the line.

"Dr. Larsen, you have to come quick!" she said.

Then the phone was silent for a moment before little Sara picked it up.

"Lilly had her baby out in the horse shit," Sara said. "Emma is pretty upset. Can you come?"

"You tell Emma that I am on my way and that things are going to be alright," I said.

The entire family was out in the barnyard when I arrived. The mare and the foal were both up and looked like they were doing okay. Emma had a halter and a lead rope on Lilly.

"It is all my fault," Emma said with tears streaming down her face. "I was cleaning the stall and left the gate ajar. Lilly ran past me and out the gate. She picked the dirtiest place in the barnyard, right on the pile of straw and manure from the last two weeks of stall cleaning. She laid down and popped that foal out before I could do anything."

Lilly was stepping sideways with her hind feet, bothered by the membranes still hanging out of her. About that time, the membranes came out with one big flop, and she stepped away.

I picked up the membranes and spread them out on the ground to show Emma how to check that the entire afterbirth came out.

"In cows, we don't worry too much about retained membranes these days. As long as the cow is doing okay. But in the horse, it is an entirely different story, and it is important to check that both of these ends are intact. Otherwise, we need to go in and get the retained pieces."

"Now, let's clean this little gal up and take care of her navel and her E-Se injection," I said. "Then, we can take care of Lilly."

By the time we were done and Lilly and the foal back in their stall, Emma had calmed down a little.

"What should I watch for now?" Emma asked.

"You should watch for a normal baby," I said. "Don't worry unless there is something to worry about. You have a long way to go in this life, Emma. If this little hiccup today is the worst you have to deal with, you will be a lucky young lady."

The Shock of It All

Ag had called, "We have a cow with a prolapsed uterus out in the calving pasture. Alice and I are going to bring her into the barnyard. Can you come out and get a look at her?"

I glanced at the clock. It was almost midnight. I thought to myself, "Ranchers must never sleep during calving season."

I could see the cow in my headlights as I pulled into the barnyard. Standing with Ag, short for Agnes I would guess, and Alice on each side of her, she was still a hundred yards out in the field.

I opened the gate and pulled out into the field. I stopped and closed the gate before driving out to where the cow was standing. I left the lights on when I got out of the truck.

This cow, a large Charolais, was either very tame or very sick. She did not flinch when I pulled up in front of her in the truck.

"I think she has decided this is as far as she is going to go," Ag said as I stepped out of the truck.

I could see her uterus hanging out of her, almost reaching the ground. This was always a bad sign. These cows, who have lost all inner attachments of the uterus, had a poor prognosis in my experience. I assumed that to be from the rupture of the ovarian vessels and blood loss. Even when I replaced these uteruses, over half of these cows would die.

"That uterus looks like bad news to me," I said. "Let me get a look at her before we start on putting things back together."

"She walked to this point pretty well," Alice said. "But then she just stopped. We have been standing here for fifteen minutes before you got here."

I lifted her muzzle so I could look at her oral membranes in the lights. She was ghost white.

"I think this cow is going to drop dead any minute now," I said. "That uterus has lost all its ligament attachments and is hanging out full length. She is probably bleeding inside."

"Let's try to hamburger her," Ag said, looking at Alice.

"I'm willing to help," Alice said. "If we get her gutted tonight, we can get Chuck out here in the morning with his mobile slaughter truck."

Both these ladies were in the sixties, Alice probably older. It is past midnight, and they are talking about butchering a fifteen hundred pound cow like it is just a small chore.

"What do you think, Doc?" Ag asked.

"I think it might be marginal, but the US Department of Agriculture says that a cow is fit for slaughter if she has passed her fetal membranes," I said. "This is a lot of hamburger. If the meat is no good, you can probably determine that before you process it. That way, you are only out the slaughter cost. And in this case, maybe just an extra hour of work."

"I'll go get the tractor with the frontend loader," Ag said as she started toward the barn. "I guess you can go, Doc."

"Actually, I think I'll stay and give you gals a hand," I said. "I'm a little interested in what she looks like inside."

About the time Ag returned with the tractor, the cow collapsed. Alice cut her throat, shackled her hocks, and lifted her off the ground with the tractor's loader. Then the work started. My headlights provided ample light, and they both worked with practiced repetition.

When the abdomen was opened, I expected to find a significant amount of blood. There was none. I looked at the ovarian vessels. The ovarian ligaments were ruptured, but the vessels were intact. My thinking that these stretched-out uterine prolapses resulted in substantial blood loss was just wrong. This cow died from shock.

This would change my approach to treating these cases. It is not always possible to give these cows a large volume of fluids out in a distant pasture in the middle of the night. Like in this case, many are probably too far gone by the time I get to them. But for some, a good dose of Dexamethasone and a bottle of glucose or calcium might be enough to do the trick.

***

I called in the morning just to check with Ag on the status of the carcass. My guess was that saving five hundred pounds of hamburger would not be as appealing after a night's sleep and clearer thinking.

"What did Chuck think of the cow?" I asked.

"We didn't even call him," Ag said. "Things just didn't smell right this morning. It was a good thought last night, but not so much now."

"That is probably the best," I said. "It would be one thing if you were starving to death, but when we have access to the world's best meat supply, there is little benefit in trying to salvage marginal meat."

"That is just about what Alice said," Ag said. "She wasn't hungry enough to want it."

"There was some good to come out of last night," I said. "I have always felt some of these cows died from blood loss. But there wasn't a drop of blood in that belly last night. She died from shock. That will change the way I treat these cows with a massive prolapse.

I will treat them for shock first, then worry about the prolapse. That is how we learn things. That is why they call it practice."

"Now, I just have to get the rendering truck out here before the dogs get into that gut pile," Ag said.

"That would be good. Otherwise, I will see a sick dog or two for you," I said.

Ali

I first met Ali in the fall of 1976. He was a German shepherd. At that time, almost all German shepherds were great dogs. Ali was a large, black and tan male. He was well behaved and a reliable member of the family.

Gene had called me because he was concerned that Ali was vomiting. It was troublesome more than anything else. He was not vomiting all the time, but usually a little bit every day. Gene was worried because he would bring shaker balls home from the mill, and Ali loved to retrieve those balls.

"Shaker balls are larger than a baseball," Gene explained. "I don't know if he could swallow one of those or not."

"You wouldn't think so," I said. "But, I guess we could try to see if they would show up on an x-ray."

"They probably are just fabric and maybe some rubber," Gene said.

This was before the clinic was open. I was limited to house calls and a mobile x-ray unit designed for a horse's leg rather than the abdomen of a large German shepherd.

I got a picture. Ali was a great patient. It was an awkward setup, but he tolerated it well. I had Ali lie under the x-ray unit that was suspended from a stand. Not many dogs would put up with that without some sedation. It was not a problem with Ali.

I often found that large dogs, with a lot of self-confidence, were easy to deal with in unusual situations. A hundred little dogs, who were afraid of their shadows, would be impossible to control in the same situation.

I looked and looked at the film. There was nothing I could see to suggest a ball in his stomach.

"If you are concerned, we could do exploratory surgery," I told Gene.

"Well, as long as the vomiting doesn't get worse, I guess we will just watch him for now," Gene said.

"We might try to change his diet to canned food and feed in smaller meal sizes, several times daily," I suggested. "If you get concerned, we can do surgery at any time."

The diet change almost solved the vomiting issue. Ali would now vomit only occasionally. Gene was no longer worried, and we would talk only on an occasional basis.

***

Then Easter Sunday came along. My folks were visiting us for the weekend. For whatever reason, Easter Sunday was a busy day for me for the first few years in Sweet Home. We woke up to a pickup parked in the driveway with a large sow in the back.

"Doc, I am sorry to bother you on Easter Sunday," George said. "My wife said I should come early, so I would be less disruptive to your day."

I had pulled on a pair of pants and an old shirt and went out to talk with George in my slippers.

"I picked her up at the auction on Thursday," George continued. "She can't poop, Doc."

"Give me a few minutes to get a few things and get something better on my feet, and I will get a look at her, George," I said.

Looking her over, she looked a little full in the gut. On the rectal exam, her colon was a blind pouch on the rectal side. A lot of scar tissue was present.

"Things don't go anywhere from this end," I said.

"Makes me mad that someone would send her to the sale for me to buy," George said.

"This sow apparently had a rectal prolapse at some time in her life," I said. "Her colon is scarred closed. It was probably poorly repaired when the prolapse happened. I can maybe open it with a trocar, but not without some risk, and it will not stay open for any significant amount of time."

"Ah, I had a pig once with the same problem," George said. "A long time ago."

"How did you handle her then?" I asked.

"I took her to the sale," George said.

"See there, things sort of come back around, sometimes," I said.

That brought that visit to a close. There wasn't much to be done anyway. I guess the sow could be salvaged for sausage. Probably be okay if you were hungry enough, but I wouldn't want to eat it.

***

It was well after dinner in the late afternoon when the phone rang.

"Doc, this is Gene," Gene said. "Now I know there is a shaker ball in Ali's stomach. Our daughter and her family were here for dinner. After dinner, her husband and the boys were out throwing a ball for Ali. One toss was high in the air, he jumped up and caught it, and they watched it go down. He is sort of uncomfortable now."

"My day tomorrow is already a disaster," I said. "Maybe I should meet you at the clinic and plan to do the surgery this evening."

"I don't want to disrupt your holiday," Gene said. "But if you are willing to do it this evening, that would be great."

"I can meet you at the clinic in half an hour," I said.

Hanging up the phone, I turned to Dad. "Are you interested in going to the clinic and watching a surgery?"

Dad, Sandy, I head to the clinic, leaving Mom home with the kids. Mom probably would have liked to go, but staying with the kids made up for her missing the surgery.

We were pretty well set up by the time Gene came through the door with Ali. We rushed through the check-in process and moved ahead with the exploratory surgery.

With Ali under anesthesia and prepped for surgery, I made a five-inch incision on his ventral abdominal midline. It only took me a few minutes to be in the abdomen. I reached in with one hand and palpated the stomach. There was a large shaker ball, and then I felt the second ball also. Gene had been correct in his assumption all along.

I was able to externalize the stomach with both balls.

With both large balls and just a small portion of the stomach hanging outside of the incision, Dad made his observation. "That looks just like a large scrotum and a couple of big balls."

I made an incision in the stomach just long enough to allow me to express one of the shaker balls out. The first one, then the next came. The second ball had obviously been in the stomach for some time. All the fuzz on the surface was gone. That was the ball that had been there since last fall.

I turned the gas off as I started to close. At that time, I had a Metaphane gas machine. Metaphane was an excellent anesthetic gas, but it had a prolonged recovery period. By turning the gas off early, Ali should be awake when we were cleaned up and ready to go.

The closure was simple. I used a double layer inverting closure on the stomach incision, returned the stomach to its normal position. Then I used a routine 3 layer closure for the abdominal incision.

Ali was awake in no time after returning him to his kennel. We cleaned up the surgery room, and I gave Gene a call to let him know things had gone well, and Ali was awake and doing well. And he was happy about the fact that he did, in fact, have two balls in his stomach.

We started out the front door when for some unknown reason, I stopped and said to Sandy, "We have a lot of money in that cash drawer. Maybe we should take it home."

We had been in the practice of only making occasional bank deposits. I went to the cash drawer and took all the bills, I think about thirteen hundred dollars. I left the change, which could have been close to twenty dollars.

***

About five in the morning, the police called. Someone had broken out the glass in the clinic door and broke in. I got up and went to the clinic to go through it with the police. The only thing that was missing was the change out of the cash drawer.

The Upgrade

"You need to hurry. Your flight is boarding now," the airline attendant said as he took our bags. "If your bags don't make your flight, they will be on the next flight. You have a full plane. There is a Ducks game in Berkeley Saturday."

We hurried down the concourse to the plane. Just what I wanted, to ride to San Francisco with a planeload of Duck fans.

We squeezed down the aisle and found our seats. Now we could relax for a few minutes before the plane takes off. This was going to be our first long weekend off for nearly two years. A continuing education trip on paper, but a mini-vacation if we could make it such.

"I think we would have been better off to take the extra time and drive to Portland," Sandy said. "Then we could have got a direct flight to Reno. I hate changing planes, and especially in San Francisco."

"It won't be too bad. We should have plenty of time," I said.

About then, we were rudely made aware that our flight was going to be anything but pleasant. Sitting behind us, and on top of our seats at times, was a most unruly four-year-old and his mother, who had no concept of discipline.

We are making the final approach to landing on the runway that extends out into the bay.

"I hate landing at this airport," I say. "The first time I flew on a commercial airline was when I joined the Army. They loaded us on a plane in Portland and flew us to San Francisco. I had a window seat, and when we were landing, all I saw under the plane was water. We were getting closer and closer to the water. I was lifting my feet before the ground and a runway came into my view. I repeat that episode in my mind every time I land here."

We deplane and rush down the concourse, looking for the gate for our flight to Reno. We ask an agent at the end of the hall.

"That is a separate terminal. You catch a shuttle bus down those stairs," the agent says, pointing to a stairwell at the end of the concourse.

We hurry down the stairs and catch a bus to the detached terminal. Then we load into a puddle jumper, not my idea of a fun flight. I am white-knuckled all the way to Reno. We arrive, and Sandy's bag makes the flight. My bag is nowhere in sight. We leave our information and hail a cab to the hotel.

"We have your reservation right here," the hotel clerk says to Sandy. "It is a nice room. I hope you enjoy your stay."

Sandy looks over the paperwork while I twiddle my thumbs.

"Is this a non-smoking room?" Sandy asks.

"No, Ma'am," The clerk responds. "This is a smoking room."

"We requested a non-smoking room on our reservation," Sandy says.

The clerk looks at his computer screen closely. "I see that you are correct. It says a non-smoking room right here," the clerk says. "We don't have a non-smoking room available in this room class."

It looks like another planeload of people has arrived. There is quite a line behind us now.

"I have to have a non-smoking room," Sandy says.

"Let me go talk with my supervisor," the clerk says as he leaves his station.

The people behind us let out an audible moan. Sandy is unwavering.

Finally, after close to five minutes, the clerk returns. He is all smiles.

"I have an upgraded room for you," he says, winking at me. "You guys are really going to enjoy this room! It is one of our best suites."

The bellhop leads us away. The room is high in the hotel, on the 35th floor.

"You are going to really enjoy this room," he said as he pushed open the door.

He set the bags down and went to the drapes, and pulled them open. The entire wall is floor to ceiling windows, and the view of the city is incredible. I feel a little embarrassed as I hand him a five-dollar tip.

"Can you believe this room," I said to Sandy. "And all because you wouldn't accept a smoking room."

The suite's main room is three times the size of any hotel room we have ever seen. The bathroom is enormous. It has a large shower with two shower heads. There is a large jacuzzi tub, a massive mirror with double sinks, and a separate water closet.

"This sort of reminds me of Ma and Pa Kettle," I said." We're just a couple of old country bumpkins in a high-class hotel."

Sandy laughs as she investigates the kitchenette/bar area. There is a large sectional, a loveseat, and a couple of chairs. And then the bed takes up the far end of the room.

The bed is more substantial than a king-size bed and round on a raised platform. There is a thirty-inch high wrought iron railing around half of the platform. And a massive round mirror is on the ceiling above the bed.

"I am not sure how this is going to work out for us," I said.

I'm a stomach sleeper, and I hang my feet over the end of the bed. Or I sleep on my side, in touch with the edge of the bed. I am not sure I am going to be able to find either in this bed.

"I think we are maybe past the mirror stage in our relationship," Sandy said. "This could be an interesting evening."

We were just ready to leave to get a bite to eat when there was a knock at the door. It was the bellhop with my bag from the airport.

"They delivered your bag, but it looks like it has been broken into. You might want to check it carefully and make sure you file a report with the airline," he said.

"Thanks," I said. "Before you go, can you tell me something about this room? What does a night in this room usually cost?"

"This is our special suite," he said. "We generally use it to comp the high rollers. We don't rent it out very often, but when we do, it goes for twelve hundred dollars a night."

"I would guess you generally get more than a five-dollar tip up here," I said.

"I have got some awesome tips in this room, but it is not a big deal. Five dollars is a pretty standard tip in most rooms."

"I would have to have a whole lot of expendable cash before I could bring myself to pay twelve hundred dollars for a room," I say as I hand him a twenty-dollar bill.

It was sort of like adding the final insult to the plane trip. The bag was a mess, but the only thing missing is my sports coat. This gives me an excellent excuse for dressing casually. That fits my style just fine.

When the evening was over, and we are ready to go to sleep, Sandy spends a lot of time closing the drapes. It is no small task. I can't convince her that nobody can peek into a room on the thirty-fifth floor, especially if the lights are out. But she does not listen.

The bed is comfortable, but, like many hotel beds, the sheets and blankets are excessive. I go around and untuck all the sheets on my side of the bed. Then I discard the comforter and half the blankets. I crawl into bed.

I am instantly miserable. I can't find the edge of the bed, and when I reach a point where I can hang my feet over the end of the bed, my nose is at Sandy's knees. I toss and turn and get tangled up in the top sheet. I get up and pull the top sheet off the bed. Since Sandy was sleeping soundly, I open the drapes and enjoy the view until I drift off to a fitful sleep.

About three in the morning, I get up to go to the bathroom. I roll out of bed and start in the direction of the bathroom. I follow the edge of the bed until I reach the point where the round was turning toward Sandy's side of the bed. I strike out toward the bathroom.

I forgot about the railing. Just as my left foot takes a step down, the end of the railing hits me in the groin. My right leg impacts the railing. I lose my balance and fall, left side first, the two steps to the floor.

I roll onto my back. I feel like I have just been struck with a Klingon pain stick. I look around. The view out the windows is just as good from the floor. Then I look up, there I am, in full view, in the mirror.

Morning comes, Sandy is well-rested. I look like I have been wrestling steers all night. We shower together and get dressed so we can get breakfast before classes started.

As we leave the room, Sandy stops and looks at the bed, what a mess. There are piles of sheets on each side of the bed. The blankets are knotted in a heap in the middle of the bed. Even the bottom sheet is untucked and only half covered the mattress.

"The housekeeping girls are going to tell stories about what went on in that bed last night," she said.

Granny's Instructions

I struggled to secure the cow with a rope tied to a fence post and a sideline to hold her still enough for me to do a pregnancy exam. I looked over this place while I was working.

There is a big house, too fancy for my style, and a large guest house. All the driveways are paved. The barn is almost falling down, and this corral is the only thing for handling cows. There is no crowding alley and no squeeze chute.

When I was growing up, any spare money went into the barn first. The house was meant for living in, not to be a showplace. All the women in the extended family complained, but nothing changed.

I finally had this cow secured. The pregnancy exam only took a moment. My first boss would have said to linger slightly, so the client felt he got his money's worth. My left hand ran into feet and a nose before I was up to my elbow in the rectum.

"She is pregnant, probably about seven months along," I said. "But it is difficult to be accurate in the third trimester. I would say plus or minus a month. If I do a pregnancy exam between 40 and 90 days, I can be very accurate."

"I just wanted to know if she was going to calve," Howard said. "All my other cows have calved, and it is still going to be a couple of months before she calves?"

"Did she have problems calving last year?" I asked as I released the cow.

"She had retained membranes," Howard said. "I had problems getting someone to take care of her. The first vet I called, a young guy like you, looked at her but wouldn't clean her. He said it is better not to do that. So anyway I had to call an older guy out of Albany. He cleaned her, and said she should be fine."

"You know, things are always changing in medicine and veterinary medicine," I said. "Treating a cow with retained fetal membranes is one of those things that have changed. We now know that manually removing those membranes does more harm than good unless they are loose and just need a little tug. Had you gone with the first recommendation, she would have had some breeding issues, but nothing like this."

"The old guys have been cleaning cows forever," Howard said. "You young guys come out here and think you know everything."

"The proof is in the pudding, my grandfather always said," I said. "One cow doesn't prove much, but all the research says, treat the cow, remove the membranes if they are loose, but never manually remove the membranes. Had you called me last year, I would have told you the same thing."

"So what do you think I should do now," Howard said.

"She will never recover the lost time," I said. "You will have to have a separate calving season just for her, or you will need to hold her over a year to get her back onto the herd schedule. I would sell her, let someone else fit her into their herd."

"You might have a point there," Howard said. "I'll have to give it some thought. You probably have a bunch of recommendations to make about how to run this place."

"I have some standard recommendations to help ranchers shorten the calving season and improve their herds," I said. "Most of those recommendations require working the entire herd once or twice a year. To do that, the first thing you need to do is upgrade this corral. You need a squeeze chute and a crowding alley."

"You expect me to spend a thousand dollars before you even get started," Howard said. "I don't think so."

"That's fine, but I won't be much help to you then," I said. "Most of those upgrades will only make your life easier. And there is no way to work a herd of cows on the end of a rope."

It was my guess that I would not be back at Howard's place any time soon. He will have to have some wreck before he calls again. And then he will really be pissed when I decline his herd.

Retained membranes remained a thorn in my side for several years. The older veterinarians in the valley continued to clean cows. My recommendations were unyielding but also often taken with a grain of salt. I figured it would be that way until I had some grey hair show up.

Then, just when I thought there was only one way to do things, Mrs. Guerin called.

"I have a heifer in the barn that needs to be cleaned," Mrs. Guerin said to Judy. "My husband left her in a small pen in the barn. The doctor can take care of her and then come to the house, and I will pay him."

"When did she calve?" Judy asked. "Doctor Larsen doesn't like to look at these cows until at least three days after calving."

"She calved yesterday," Mrs. Guerin said. "I want her taken care of now."

When I pulled into the driveway, I noticed that this was an old place. The house was old, and the barn was old. But according to directions, I pulled up to the barn and had no trouble finding the heifer. She had a small calf by her side. There was probably little chance that these membranes were loose.

The heifer was almost tame, and I had no problem getting her tied up and doing an exam. I was able to remove some of the membranes, but the bulk of the mass would not budge for the most part. I instilled 5 grams of tetracycline powder into the uterus. I gave the heifer some long-acting sulfa boluses that would give her five days of therapy.

After cleaning up, I pulled the truck over to the house as Mrs. Guerin had instructed. I knocked on the door.

Mrs. Guerin opened the door. This lady could have passed for Granny on the Beverly Hillbillies. Her grey hair was tied in a bun, her wire-rimmed glasses sort of balanced on her thin nose.

"Good afternoon," I said. "I am Dr. Larsen. I just took care of your heifer in the barn."

"Did you get her cleaned?" Mrs. Guerin asked bluntly.

"Well, I treated her with antibiotics, both in her uterus and orally," I explained. "She will do much better if we leave those membranes to come out on their own in a few days."

"You mean you didn't clean her out," she said.

"The current thinking is that it is better to allow those membranes to come out on their own," I explained. "These heifers will breed back a lot better that way. If we manually remove those membranes, there is enough damage to the uterus that it adversely affects the fertility of the cow."

Mrs. Guerin listened carefully to my explanation.

"That's okay, then, if you don't want to clean her," she said. "I will just have my husband shoot her when he gets home. I won't have a sick cow on the place."

I think this old lady just nailed me and my treatment philosophy to the wall.

"Okay, we don't have to shoot her," I said. "I'll go and clean her out. She'll be fine."

So back to the barn I went. This little heifer became the only cow that I manually removed membranes. I found it a difficult task, peeling the membrane attachment from the individual cotyledons, those 'buttons' in the bovine uterus to which the placenta attaches. I just hoped that she would get pregnant this summer.

Peanut Digger

Dixie opened the exam room door to check on Bill and Peanut Digger. We moved them into an exam room as soon as they came into the reception room.

Bill was a large man. He carried a few extra pounds on his massive frame, but his muscle mass had served him well in a lifetime of hard work. He was older, approaching retirement, and his hair was graying in a short crewcut.

Peanut Digger was a Brittany spaniel. Brittanys are the most hyper of all the spaniel breeds. And Peanut Digger was the most hyper example of a Brittany spaniel that I knew. He bounced off the walls the entire time he was in the clinic. If they were left in the reception room, there was total chaos by the time they were called into an exam room.

Bill was totally aloof to the chaos. He would just sit there, arms folded across his chest and feet extended and crossed at the ankles.

Peanut Digger would continuously circle the exam room. Jump up with his front feet on the counter to check out the items there.

Tongue out and panting, with saliva dripping from the corners of his mouth.

Peanut Digger would only slightly slow down when placed on the exam table for an exam or treatment. He was a good dog, he was just absolutely unable to calm himself.

"You know, Bill, we might be able to calm this guy if we neutered him," I said.

"Neuter him! No way, in fact, I'm planning on raising a litter or two," Bill replied. "I bought a female Brittany a couple of weeks ago. She is already in heat."

"You might have your hands full with a bunch of puppies running around," I said. "What are you going to do if you can't get rid of them?"

"Brittanys are pretty popular dogs," Bill said. "I don't think I will have any problems."

***

The next time I saw Peanut Digger, it was to sew up a gash on his muzzle. It was a typical scene when I entered the exam room, Bill seated in the chair, and Peanut Digger going nuts. We wrestled him onto the exam table, and I looked at a deep wound on the left side of his muzzle.

"They don't get along so well," Bill said. "I think she wasn't quite ready for his attention. She sure surprised him. I hope it didn't ruin their relationship."

"This is a deep wound," I said. "We will have to sedate Peanut Digger to get it cleaned up and closed. I don't think he will hold still for it any other way."

"I don't think he will hold still for anything," Bill said.

"If we are going to sedate him, I will make you a deal on a neuter," I said, hoping that Bill would reconsider that option.

"Oh no, we are going to get this litter of pups even if we have to resort to artificial insemination," Bill said.

Artificial insemination in the dog was one of my worst nightmares. The problem was collecting the semen from the dog. People requested the procedure, usually because they could not get the dogs to breed naturally. So then they bring them to a vet clinic, a real relaxing environment for most dogs, and expect someone to collect the male dog via some form of masturbation. In my experience, it just didn't often work. And with Peanut Digger, I could not imagine getting it done.

"I don't think AI would be an option with Peanut Digger," I said. "We will sew up this wound, then you keep them apart. Put them together once or twice a day, with some supervision. Maybe have her on a leash. When she finally accepts his advances, you breed them every other day for as long as she will accept him. That usually results in a pregnancy."

It was no small feat, getting an IV catheter into Peanut Digger. But once that was done, sedating him was no problem. We gave him some IV pentothol and some gas via a mask. I shaved the wound with a straight razor. Wound healing in animals requires a close shave of the wound edge. If you can do nothing else, shaving the wound will do a world of good. After scrubbing the wound, I closed the deep tissues with a continuous suture of Dexon and then sutured the skin with nylon.

***

A couple of weeks later, Bill was back with Peanut Digger to get the sutures out. The wound had healed well. There will be no scar once the hair grows back. Trying to get the stitches out was something else again. Getting the hook of the suture scissors under a suture was little like hunting birds. You had to anticipate where his nose was going be because holding him still was impossible. But we got the job done.

"The wound is well healed," I said. "Did you ever get him hooked up with his girlfriend?"

"Sure did, several times," Bill said. "We should have pups in another six weeks or so."

"Give me a call if you have problems," I said. "Most of these dogs have puppies with no problems."

***

A few months later, I noticed Bill's name was in the last slot on the appointment book.

"I don't know if I'm up to Peanut Digger this afternoon," I said to Dixie. "I'm already worn out."

"You're in luck. It is the litter of puppies for vaccinations," Dixie said. "That shouldn't be any problem. But he said there were seven puppies."

I opened the exam room door to the most unbelievable commotion. There was Bill, seated in the chair as usual. And then there were seven Brittany pups, all males, running wild around the room. Seven male puppies, what are the odds of that. And every one of them was an exact replica of Peanut Digger.

Peanut Digger times seven shouldn't be any problem.

Toby's Sore Eye

"Doc, Toby has been reluctant to work the last few days, nothing I could put the finger on but just lying around. Then this morning, I wake up, and his left eye is swollen and really painful. I would sure like you to get a look at him the first thing if that is possible."

"I know my morning is hectic today, Bob," I said. "But if you have him at the clinic at seven-thirty, I'll get around and meet you then."

\|\|*\|*

Toby flinched and whined when I picked him up to put him on the exam table. His left eye was noticeably swollen and must be extremely painful. Any movement seemed to be uncomfortable for him.

"Boy, I don't see many eyes that are this painful," I said as I started an exam of Toby.

"Aren't you going to look at the eye?" Bob asked.

"I'll look at the eye after I look at Toby. Otherwise, we could end up putting him under anesthesia when he is in heart failure or something like that. I want to make sure I don't overlook something else."

Finally, I carefully grasped Toby by the muzzle with my left hand. I could feel him try to pull away as I elevated his upper eyelid to start looking at the eye. I looked under both eyelids to make sure there was no foreign body. I attempted to elevate his third eyelid in the corner of his eye. But, Toby could not tolerate that much intervention. He recoiled and growled this time.

The pupil of the eye was constricted to a pinpoint due to the pain. I placed a couple of drops of medication in the eye to dilate the pupil.

"Do you know of anything that could have injured this eye?" I asked Bob.

"No, he was in the house last night, laying around a little like I said, but doing fine otherwise," Bob said. "Then this morning, he is really painful."

With the pupil dilated, I carefully held Toby by the chin and looked at the back of his eye with my ophthalmoscope, starting from a distance and slowly moving in close to his face to avoid alarming him.

It took me a couple of seconds to figure out what I was looking at. Projecting from the back of the eye, near the optic disc (that spot where the optic nerve connects with the retina), was a porcupine quill. It was probably penetrating over a centimeter into the eye.

"Bob, has Toby ever had porcupine quills?" I asked.

"Oh, yes," Bob said. "We had quite a struggle pulling a mouthful of quills out of him a couple of months ago."

"Well, it looks like you must have missed one," I said. "Toby has a quill that has poked through the back of his eye. I think this one is for the books. I always hear stories of quills that are left behind migrating to odd places. But I've never read of one puncturing the eye from behind."

"Can you get it out?" Bob asked.

"I can't, and I don't know if a specialist could or not," I replied. "Getting to that area behind the eye would be extremely difficult. And after removing the quill, you might lose the eye anyway."

"I don't think a specialist is in the cards for Toby," Bob said. "I don't have cows anymore, and old Toby doesn't have much to do. He just lies around the house and chases a stick or two once in a while."

"In that case, I think the best thing to do is to remove the eye," I said. "He will do fine with one eye, and we will get rid of the pain."

"Is that something you can do?" Bob asked.

"That is a common surgery for us, Bob. It is not a surgery I do every week, but I do one every couple of months."

"You said you were busy today. How long is he going to have to wait before we get him fixed?" Bob asked.

"I think he is painful enough that we are justified in moving some appointments around and getting him on the surgery table a little later this morning."

"Okay, let's get it done," Bob said. "When is he going to be able to go home?"

"If everything goes well, he will be able to go home this evening. And he will be feeling much better with this painful eye gone."

With that, we set up to remove Toby's left eye. Enucleation of an eye in the dog is a relatively straightforward surgery.

Once the area was prepared for surgery, I incised around the margin of the eyelids. Then I dissected behind the conjunctiva to the surface of the globe.

I severed the ocular muscles at their attachment to the globe. This avoided a lot of bleeding. Once the globe was free of all the muscle attachments, I ligated the optic stalk (the optic nerve and the vessels supplying the inside of the eye). This was complicated in Toby by the presence of the quill.

Once the eye was removed, I closed the dense fascia over the eye socket. This way, the side of the face would remain smooth, rather than sinking back into the socket when the muscles atrophied. Once the skin was closed, you might not notice the missing eye if you didn't look twice.

After the surgery, I opened the eyeball. The quill was halfway into the eye. The track of the quill was infected. I doubt the eyeball could have been saved under any circumstances.

Late that afternoon, Toby bounced out of the clinic with Bob and the kids, acting like nothing had ever been wrong.

Wild Horses

Standing at the corral fence, we were looking at the ugliest horse that I had ever seen. She was an older roan mare that my father-in-law had just adopted from the Bureau of Land Management Wild Horse Adoption Center in Burns.

She looked like she had just stepped off a Spanish galleon, a real mustang in every sense of the word. Her massive head dominated her features. It made her appear unbalanced, almost like a trout living in a stream with little or no food source. Long hair hung from her lower jaw, making her head look larger than it was. Her hooves were large and splayed out from a lifetime on the open range. Her ribs were countable, indicating that grass on that range was sparse.

She paced up and down the far fence of the corral, uneasy with our presence. Her experiences with people had likely been unpleasant. She would rest her head on the upper rail at times as if she was trying to gauge the height. Just in case she needs to jump out.

"Are you sure you want to pregnancy check her, Jim? I asked.

"I would like to know, just so we can make some plans for taking care of a foal," Jim said.

"How do you think we are going to do that without getting killed?" I asked. "I don't think I am going to be interested in standing behind her."

"I figure we can run her in the chute," Jim said.

"I am not much of a horse doctor, but I don't think that would be a good idea," I said. "She would tear herself up in there."

"The other option is to run her into the crowding alley and throw a rope on her, and you can check her reaching over the fence," Jim said.

"We can try it that way," I said. "We might get lucky."

Jim opened the gate into the alley, and she ran right in when he started over the fence on the far side of the corral. He closed the gate, and we pushed her up the alley toward the squeeze chute. Jim lassoed her and tied the rope to a post at the end of the alley.

All hell broke loose. When the mare realized she was tied, she fought the rope for all she was worth. She pulled back, throwing her feet in all directions. Her front hooves seemed to reach the top rail of the alley walls. This fight went on for a surprisingly long time before she choked herself enough to settle down.

The walkway on the outside of the alley fence allowed me easy access to her. The major problem was it was on her left side with her head to my left. That meant that I would have to check her with my right hand. I was almost blind with my right hand rectally. That was sort of a funny thing. I trained myself to do rectal exams with my left hand, leaving my right free for any other tasks that may be needed. I could almost see with my fingertips of my left hand, not so with my right hand.

With a lubed plastic sleeve on my right hand, I took a deep breath and leaned over the upper rail. The old mare had decided that she was caught; she did not move as I inserted my hand into her rectum. I advanced my arm halfway to my elbow, and my hand bumped right into a foal. Pregnant for sure, I swept my hand over the fetal head and feet. I would guess close to six months. Good enough for family work.

"Jim, she is about six months pregnant," I said as I pulled my arm out and stood up, thankful for being in one piece. "That is a rough estimate, but pregnant for sure. Now we just have to get that rope off her."

I had never been more thankful for a quick-release honda. Getting this rope off this mare without a quick-release would be difficult, if not dangerous. I grabbed the short leather thong on the quick-release and gave it a good pull. Then I quickly ducked as the rope flew when the mare threw her head up and quickly backed out of the alley.

I was thankful that it was over. It was a little sad that she was not the only wild horse I would have to deal with in those early years. When BLM started adopting the wild horses gathered from the Eastern Oregon rangelands, it seemed everyone wanted a free horse.

***

I slowed the truck to a crawl as I made a couple of the sharp corners on Old Holley Road. I looked carefully for the driveway to the place on the corner. They had called to have a horse castrated.

A small group of people was in the pasture behind the barn, standing and talking while watching me pull my truck into the pasture.

"Is this where I am supposed to geld a horse?" I asked after I rolled down my window.

"Yes, this is the place, Doc," Ed said.

"Where is the horse?" I asked.

"He is in the shed there," Ed said, pointing to a small shed behind the barn.

"Well, let's get him out here. This pasture looks like a good spot to do the surgery," I said.

"Can't do that, Doc," Ed said. "He is sort of wild. I don't think he has ever had a hand laid on him. We don't have any facilities for handling a wild horse. We just offloaded him into that shed, and that has been his home for the last few days."

Great, I thought. Now I get the rest of the story. I don't know what they expect me to do with a wild horse, free in a pen, and has never been touched by man.

I entered the shed. There was a large gray stallion, cautiously eating hay out of the feed rack. He glanced at me but did not seem concerned at my presence. My guess was he had been around people at some point in his life. Many wild horses on the range have gone wild from some of the ranches in the area.

Just the week before this call, I had read an article in Veterinary Medicine, a minor professional journal. This article was about sedating a horse in this circumstance. They suggested using a full ten ccs of acepromazine, a popular tranquilizer, in a syringe, squirted into the horse's mouth. There would be rapid absorption through the oral mucus membranes, plus whatever he swallowed. It just might be the ticket with this guy.

Ed came over to the truck when I returned for my rope and a syringe full of acepromazine.

"What do you think, Doc?" Ed asked. "Are you going to be able to handle this guy?"

"My first thought was just to leave," I said. "You need to be a little more forthcoming with information when you schedule an appointment. But I do have a trick to try on this guy. If it works, we can maybe get the job done. If it doesn't, you're on your own."

I returned to the shed and offered the stallion a handful of grain. He quickly nibbled at it and nuzzled my hand for more.

"You know what grain is and where it comes from," I said to the horse who was still nuzzling my hand.

I took another handful of grain and held out my hand for him, making him stretch his head through the feed rack to reach the grain. As he ate, I slipped the syringe into the corner of his mouth. He did not object. Then with a hard push, I shot the entire dose into his mouth. He reared and pulled his head out of the feed rack, getting a good gash on the top of his head.

The reaction was rapid. Within a minute, his head was starting to hang down. Giving him a little more time for the Ace to work, I returned to the truck and got everything ready to anesthetize him and do the surgery. Then I threw a rope around his neck and led a staggering horse out into the pasture.

Surgery was a breeze. Less than 2 grams of pentothol was needed to put him under anesthesia. I laid him on his right side and pulled his left leg out of the way with a sideline. He was four or five years old, and his testicles needed a strong pull to break the cremaster muscles, but other than that, it was a routine castration. I removed the bottom of the scrotum and stretched the incision to allow for adequate drainage. Then I gave an injection of long-acting penicillin and his tetanus vaccination. I didn't want any complications with this guy.

"Doc, how would you recommend we tame this guy down?" Ed asked as we waited for him to recover.

"I would get a halter on him and a long chain to a heavy tire or something he can drag around the pasture for a few days," I said. "This guy is not completely wild. He knows what grain is, and with a little work, he will tame right down. If you work with him several times a day, you will have him tamed down quickly. Then you can figure out which one of you are going to try to ride him."

***

That proved to be good advice. The horse was dragging the wheel around the pasture for a week or two, and then Ed had him eating out of his hand. He became an excellent horse for them.

Uterine Twist, Which Way Do We Turn?

I ran my hand into Bertha's vagina a second time. It still ran into a blind pouch. Bertha was a prized Jersey cow that supplied milk to a lot of neighbors.

"What the heck is going on?" I asked myself. I had never had a dystocia in a Jersey before unless it was associated with milk fever.

I explored the pouch with my fingertips. Then the light finally flashed on. This was a full uterine torsion. Partial torsions were common. In fact, I prided myself at being able to untwist a uterus that was half rotated.

I used my left arm's strength, I would rock the calf a little, and then, with a strong flip, I would turn it upright. The narrowed twisted vagina would open completely, and delivery would be a snap after the correction. This was a complete 360-degree torsion. The vagina was twisted closed like the top of a plastic bag. I tried to advance my hand through the twisted vagina, to no avail.

My thought was to get my hand into the uterus with a detorsion rod and hook the calf's feet to the rod. Then with a bar through the other end of the detorsion rod, I could untwist the uterus with a strong crank. But that wasn't going to happen. I could not begin to advance my hand through the twisted vagina.

"Carol, there is full 360-degree uterine torsion," I said. "I can't get my hand through it. That means we are probably going to have to do a C-section."

There was a gathering crowd in this small backyard barn lot. It seemed that half of Crawfordsville was watching.

"Is that the only option?" John asked.

I started to reply, but the question had started the wheels turning in my memory bank.

"I am sort of short on tricks," I said. "But there is one that we could try. I have never done it. In fact, I have never seen it done. It might be worth a try. I will need a two by twelve plank, about twelve feet long."

"We just happen to have one of those," Carol said. "Over in that lumber pile."

A couple of guys pulled the plank out of the lumber pile and had it beside Bertha in no time. I had everyone's full attention now. Nobody had any idea what I was up to.

"This is the plan," I explained. "We lay Bertha on her side, lay this plank across her belly, with the plank's midpoint on her belly. Then we roll Bertha to her other side while some brave soul stands on the plank. The plank holds the calf while Bertha turns, thus undoing the uterine torsion. The only trick is to make sure you roll her the right way."

"And just how are you going to lay her down on her side?" Bill asked. "I suppose you just ask her."

"That's another trick that I use all the time," I said. "It's called the flying W. If you haven't seen it, you will be impressed."

I got my large cotton rope and placed the middle of the rope over Bertha's neck. I crossed the rope between her front legs and brought it up each side, crossing again in the middle of her back. Then I bring both ends out between her hind legs, on each side of her udder, the application was complete. A slight pull, and Bertha fell to her right side.

"I'll be darned," Bill said.

I positioned the plank across Bertha's belly, with the midpoint in the middle of Bertha's belly. This would be enough plank to make a complete turn for Bertha. The plank was at about a 45-degree angle with the ground. It might take an agile person to ride it for the entire arc.

I looked around at the crowd.

"I can stand on the plank," Carol said. "She is my cow, and there was a day that I was somewhat of a gymnast."

I positioned Carol on the plank, about four feet up the plank from the ground. I had a couple of guys on each rope tied to both the front and hind feet.

"Now, we are going to go very slow," I said. "I need to have my hand in her vagina to make sure we are turning the correct way. I tend to be a little dyslexic, and I have trouble figuring this out."

With my hand in the vagina, I had the guys start lifting on the feet. Sure enough, the twist was tightening.

"Okay, all stop," I said. "We are going the wrong way. We have to start all over with Bertha on her left side."

It only took a couple of minutes to untie Bertha's feet and remove the plank. I didn't have to do much. The whole crew knew what was up and what needed to be done.

With Bertha on her feet, Bill quickly grabbed the ends of the ropes on the flying W. He wanted to feel just how easy it worked.

"Now, we want her to fall on the left side," I said. "So when you pull, you want to lean left and put all the pressure in that direction."

Bill pulled, leaned left, and Bertha flopped to her left side. Bill had a big smile on his face.

"That was so easy, I can't believe it," Bill said.

"If you are throwing a big bull or an ornery steer, it might take a couple of guys on each rope," I said. "But I have never seen it fail."

The rest of the crew had Bertha's feet tied and the plank in place in no time. Carol jumped on the plank, and we rolled Bertha.

After standing Bertha up, I washed her up one more time. I ran my hand into a normal birth canal. I didn't let on, but I was almost as amazed as was the crowd watching. I grabbed both front feet of the calf and pulled them into the birth canal. As I turned to my bucket for my OB straps, Bertha strained and out popped the head. One more strain, and both John and I caught the calf before it fell to the ground.

"That was easy," John said.

"Jersey cows have the easiest deliveries of all the breeds," I said.

We turned Bertha loose, and she turned her attention to the little heifer calf, utterly oblivious to the crowd watching.

One Bite Deserves Another

"Doc, I know it's late, but Rex's hurt pretty bad," Reese said into the phone.

"What's going on with him, Reese?" I asked.

Reese was an old rancher. He still worked a few cows, but his son did most of the hard work now. Reese was a big man, big all over. And his features were rough. He shaved about once a week, and wrinkles seemed to soften a once stern face.

"I was working a few cows tonight," Reese said. "Old Rex, if they don't do what he wants, he gets sort of mean. Rex grabbed this old black cow by her heel, and she kicked at him pretty hard. Old Rex, he ain't smart enough to let go. She broke his mouth up pretty good."

"Okay, Reese," I said. "I can meet you at the clinic in a few minutes."

I could see that Rex had a significant injury to his mouth when they came through the door. His mouth hung open, and the right side of his muzzle drooped lower than it should.

Rex was one of my working dogs. He mainly was Australian shepherd but darker in color than was typical for that breed. He also was a little more aggressive with the cows than most dogs.

"He really did it to himself this time," Reese said.

I lifted Rex onto the exam table. I always marveled at the bodies of these working dogs. Some would be considered heavy if you just looked at their weight. But these dogs were solid muscle. And tough as nails. Rex did not act like his injury was bothering him at all.

I lifted his right lip; what a mess. He had a fracture of his maxilla, his upper jaw bone. He must have had a hold on the cow's foot when she kicked. His canine tooth acted like a lever, and it produced a flap of bone that contained his canine tooth, two incisor teeth, and two premolars. The fracture line ran down the right-center on the roof of his mouth.

"This is a mess," I said. "But it is probably good that you brought Rex in tonight. I think I can fix it with a couple of pins and several wires. He is going to be uncomfortable for a couple of months."

"Do you think he is going to lose any teeth?" Reese asked.

"All these teeth are still set in bone," I said. "It is the bone that is broken. We might need a little luck here. But I think this will repair okay. The problem is I am going to have a couple of wires running across the roof of his mouth, and that is going to bother him."

"How long will he have those wires?" Reese asked.

"Probably six to eight weeks," I said. "Now that I think about it, I will probably wire between a couple of sets of teeth also. Sometimes when we do that, we will lose a tooth or two. But that won't bother him."

"He doesn't have to smile for any pictures," Reese said. "He won't care as long as he can get back to work."

"I think it is a good thing that you brought him in tonight," I said. "With this wound open on the roof of his mouth, it would be really contaminated by morning."

"Are you planning to fix it tonight, Doc?" Reese asked.

"I think that it is something I can do by myself," I said. "And he will be far better off if it is repaired tonight. I would guess he will be ready to go home in the morning."

With that, I showed Reese out the door after I had him hold Rex while I placed an IV catheter and started a bag of fluids on a slow drip. I got things set up in surgery and then gave Rex a dose of pentothal via the IV.

The repair went pretty well. The slab of bone with the five teeth fit snuggly into place. I secured it with a couple of pins that ran through to the other side of the mouth and then placed tension band wires that ran across the roof of the mouth to hold the slab of bone securely in place.

That was all I needed. I did wire the teeth on each end of the slab to their neighboring teeth. And then, I covered all the sharp ends of the pins and wires with dental acrylic. Hopefully, the acrylic will last for the entire eight weeks.

Rex went home the following morning, looking none the worse for wear. I checked him every couple of weeks. The repair held up well, and at eight weeks, his x-rays showed good healing.

We sedated Rex and pulled the pins and wires. There were a couple of minor abrasions in his mouth, but nothing that would not heal.

"I think you can probably start treating Rex like a dog again," I said when Reese was in to pick up Rex.

"I think that he can't wait to get back to work," Reese said. "I bet he is thinking he is going to get back at that old black cow."

"I don't know," I said. "Maybe we should put a muzzle on him for a couple of weeks. I'm not sure this repair will hold up to another bite like the last one."

"We ain't going to make no sissy out of Rex," Reese said.

***

It was a long week later that Reese had Rex back in the clinic. He had landed a bite on the heel of that same old black cow, and sure enough, the same slab of maxillary bone was hanging loose.

"Looks like we need to do the same thing again," I said. "But this time, we are going to extract both of those upper canine teeth. That way, when he bites, he won't be able to hang on, and there won't be a big lever to break that jaw."

"That sounds good, Doc," Reese said. "Because we ain't going to slow old Rex down."

Rex healed well, once again, and didn't seem to notice his missing canine teeth.

The Perfect Shot

I thumbed through the journal, scanning the pages for something interesting before discarding it onto the pile on my desk. I seldom read anything cover to cover, but I could rapidly review the pages and pick out any interesting articles. Those articles I could quickly scan a little closer in a manner that would allow me to retrieve them later if needed.

This journal was a lesser journal, titled Veterinary Medicine. It occasionally had some practical articles from actual practitioners, not just a bunch of university types who had no fundamental concept of what private practice was like.

The article that caught my eye today was on a do-it-yourself blowgun and dart that you could use to get a capture drug into an otherwise wild critter. Lord knows I seem to have my share of those.

It was relatively easy to make. Using a four-foot piece of PCV pipe for the blowgun and darts fashioned from a couple of three cc syringes. The blowgun was nothing to make. The dart took a little time.

The flanges on the back of the syringe were trimmed off, so you had a smooth barrel. One plunger stopper was inserted into the barrel of the syringe. It would deliver the dose when the syringe was loaded, and the chamber between the plungers was charged with air.

The second plunger was used on the back of the syringe. You used silicone gel to secure a bundle of one-inch long yarn pieces to serve as the fletchings, like the feathers of an arrow. Using a bright color for the yarn may aid in finding a dart that missed the target. This plunger was secured to the syringe's back by driving a couple of twenty gauge needles through the syringe and clipping them off flush with the edge.

The journal's plan called for a sixteen gauge needle, plugged with superglue at the end, and a side port in the needle. This port was made with a small file close to the end. I found that to be an unnecessary step. I just used a standard sixteen gauge needle that was one and a half inches long. Plugs for this needle were fashioned from a strip of silicone gel pushed out and allowed to dry in a strip. Cutting a quarter-inch piece and inserting it on the end of the needle would plug the hole, and then it would be pushed up on the needle when the target was struck. This would allow the drug to be delivered.

With a syringe and needle inserted into the chamber between the plungers, you could control the free plunger in the barrel of the syringe dart. Inject air, drive the free plunger to the end of the dart. Insert the dart's needle into the drug and withdraw air from the chamber, thus drawing the drug's dose into the dart. Then plug the dart's needle with a silicone stopper, and inject air into the chamber between the plungers to charge the dart.

I fashioned several darts and even practiced with the blowgun. It worked remarkably well and had a surprising range. With a strong puff of air, I could send the dart for over 30 yards. The dart held its line in the air. The trick was in the elevation. To fly the dart 30 yards, one needed about 30 degrees of elevation. Hitting a target at that distance would take a little luck.

***

My opportunity to use this dart came only a couple of weeks later. Margery was a client with a small acreage out in the valley. She had lost her husband some years before, and now her son and his family lived with her.

"Doctor, my grandson bought this cow and turned it out in the pasture," Margery said. "It is wild as can be. He can't get close to it, and it has an ugly-looking eye."

"Do you have a corral or even a smaller pasture you could run her into?" I asked.

"No, Bob never used that pasture for anything," Margery said. "It just has a fence around it. And that fence ain't much, just steel posts and barb wire."

"It doesn't sound like I could get a rope on her very easy, and if I did, I probably wouldn't have anything where I could tie her," I said.

"I could park the pickup out in the pasture," Margery said. "That would give you someplace to tie her. It might help you get close to her also. But I can guarantee you she will be running fast."

"You tie a wild cow to a pickup, and she will swing around on the end of that rope and cave in the side of your truck," I said. "Maybe both sides before she is done."

"That would not be good," Margery said. "Maybe we are just going to have to shoot her. I can't see just allowing her to suffer from an infected eye."

"There might be another option," I said. "I do have a blowgun dart that I could tranquilize her with if I managed to hit the target."

"That might be worth a try," Margery said.

"The thing that you have to understand is that if I come out and chase a cow around a pasture for an hour and don't catch her, you will still have a bill to pay," I said.

"That is only fair," Margery said. "But I would like you to try."

***

I stood at the gate with Margery's grandson, Jason. The brindle cow was at the far end of the pasture, head up and watching us.

"Jason, this is not much of a fence," I said. "If we get her really excited, she could go right through it."

"Grandma is pretty worried about her suffering with this eye," Jason said. "If we can't catch her, she is probably going to make me shoot her."

"We'll give it a try," I said. "But what about next time? You need to build corral up here in the corner so you can get her into a small place and get a hand on her."

"She wouldn't go into it," Jason said.

"You solve that problem by making a fence that funnels her into the corral," I said. "You get her used to going into it by putting the water in there and a feed rack. This time of the late summer, there is not much food value in this grass. If you had a couple cows, you would need to be feeding them already. I don't see any hay stored around here. What do you plan on feeding her this winter?"

"I guess I haven't thought about that," Jason said. "I was just thinking that I would get a calf out of her and start building a herd."

"That's a good plan, but you need to cover all those little bases," I said. "Let's go see if we can take care of this eye, and then you can come by the office, and I can give you some ideas on a corral system that won't break the bank. And I can set you up with a couple of guys who might have some extra hay."

"How do you want to catch her?" Jason asked.

"I am going out into the middle of this pasture and have you go out around her," I said. "If you can walk her down the fence line, I will try to get one of these darts into her. I have three darts loaded with a dose of Rompun. If I can hit her with one of them, we will have her."

"Grandma bought a halter to put on her when we catch her," Jason said. "She thought it would help next time."

"We can put it on her, but a corral will be what will help next time," I said.

"And she is not going to walk down that fence line. She will be running at full speed," Jason said as he started toward the far end of the pasture.

I walked out to the middle of the pasture, and the cow was watching both of us now. She was turning this way and that way, not sure which way to go. I load a dart into the end of the PCV pipe.

As Jason approached the corner where the cow was standing, she started down the fence line, picking up speed as she came. I raised the blowgun to my mouth and took a deep breath. Pointing it in the air at about a 30-degree angle, I waited as she approached what I guessed was the launch point. She was running full speed now. With one strong puff of air into the pipe, I launched the dart.

The dart flew in a high arc, the cow continued at full speed. I held my breath as I watched the arch of the dart.

Pow! The dart struck her on the side of the neck and stuck. I could not have placed it better if I had been standing by her side. I smiled as I looked at the end of the PCV pipe. Jason came running up to me.

"Wow, that was a good shot," Jason said.

"Lucky, Jason, lucky and good are two different words," I said. "You stand here with me for a couple of minutes. She will settle down and then just lie down. When that happens, I will let you run up to the gate and get my bag and my bucket."

Rompun is an excellent sedative for cows. It is not ideal because there are times when a patient will appear asleep, but they can still jump up defensively. But this cow was well sedated.

The left eye was ugly looking but did not look like a simple pink eye. There was a large ulcer on the cornea. I lifted the third eyelid with a pair of forceps. There was the problem, two large grass seed awns stuck in the corner of her eye.

I removed the grass seeds. Then I injected amoxicillin under the upper eyelid and another injection into the space behind her eye. I didn't think we would be catching this cow again any time soon.

Then I sutured the third eyelid up over the ulcer with a single mattress suture of 00 chromic catgut. That would give enough healing time, and the suture would dissolve on its own. Then we sprayed her face for flies.

By the time I had things put away, the cow was up and acting like she would be okay.

"That eye will heal just fine," I said. "We should not need to do anything more with it. Now you be sure to come to the office. I won't hit her with a dart like that in the next ten tries."

"So, should I build a corral or just buy one of those that you can set up?" Jason asked.

"It is just dollars and cents," I said. "The commercial systems are good. And they are fast and easy to set up. You could probably do it cheaper with a few post holes and some posts and lumber."

"Well, it's Grandma's money," Jason said. "It might be a lot easier if I got a few of those panels and set it up that way."

"Either way works," I said. "Sometimes, when you're a young man, it is better to do things yourself and get a feeling of accomplishment. And, Jason, you should be very careful and thoughtful when you are spending Grandma's money."

Another Witch, Another February

I turned into the old farm's long driveway off of Cochran Creek Road, north of Brownsville. I had been here only a couple of times before. The farm did have some character, with an old barn nestled up against a hillside and an old trailer not far from the barn that served as the living quarters.

Duane had lost his wife some years ago and lived by himself now. They had planned to build a house, but I think Duane was content to live in the trailer for now. He was a well-built guy whose black hair was accented by patches of gray at both temples.

This February has been particularly wet, with heavy rains almost every day. Massive dark clouds filled the sky this afternoon.

Duane had called about a cow with some sort of a prolapse. But he didn't leave any instructions where he had the cow. We stopped at the corral that was out by the main road. We had worked cows in this corral before.

"I hope she's not in this corral. It looks like it is a sea of mud from the rains," I said more to myself than to Joleen. "At least it is not freezing."

"Oh no," Joleen said as we pulled up to the corral. "That mud must be a foot deep. The good thing is there is no cow, and I don't see Duane."

"Let's go on up to the barn," I said. "I don't know if he uses it except for picture taking, but if he does, we might be undercover if these clouds decide to dump buckets on us."

At one time, it had been a functional barn. Now it was picturesque but aged almost beyond use. Himalayan briers reached high on the sides of the barn. There were a few openings through the vines that were kept open by foot traffic. There were multiple holes visible in the roof from missing shingles, and the barn wood was weathered by time to a delicate steel gray. The barn looked like it should grace a canvas in someone's living room.

Duane stepped out from under the barn's front part and waited for us in a pathway through the berry vines. The barn sat against the hill, and the slope provided enough room under the front of the barn for a small corral. At least we would be dry.

A large Santa Gertrudis cow stood in the middle of the corral. She looked less than happy at all the attention she was getting. There was nowhere for her to go in the cramped space, but the big dark red cow turned a few circles looking. I slipped a rope over her head. The only place to tie her was to the support beam in the corral center.

"I hope she doesn't pull the barn down on top of us," Joleen said as I started an exam on the old cow.

She suffered from a problem that I had often seen in these Brahman-cross breeds. As they approached the calving date, their cervix becomes enlarged and inflamed. Just this distended cervix hung from her vulva.

"This shouldn't be much of a problem to fix," I said to Duane. "But you are going to have to watch her close until she delivers."

I knew from experience that Duane was not one of those guys who called at three in the morning with a calving problem. I would have to do a closure on this vulva that would tear out easily if she goes into labor.

Joleen set out the necessary supplies to do an epidural injection for anesthesia to the vulva. I prepped a small area over her spine, where the tail joined the sacrum. The cow was standing quietly. Standing at her right rear, I grasped the tail with my left hand and palpated for the space between the bones that would allow access for the needle into the spinal canal. With a finger of my right hand on the site, I popped a needle into the space.

The cow jumped. Almost in slow motion, I watched her right leg come up and felt her hock brush my left thigh. In my younger days, I maybe could have responded to this stimulus. Now I just sort of observed the symmetry of motion. Her lower leg moved across my thigh roughly. Finally, after a brief eternity, her hoof caught my inner thigh. She extended her leg briskly.

Feeling somewhat like a golf ball that flies into the air off the club face, I am launched in a sloppy cartwheel toward the distant tangle of berry vines. The next thing I know, I'm picking myself up. Joleen, hushed and concerned, is helping me up, unhooking the grasping vines.

"You damn witch!" I say to the cow, picturing a large pile of hamburger. My thigh is throbbing. It takes no small amount of force to knock me ten or twelve feet.

I get another rope and tie the cow a little more securely. I finish the epidural injection and clean and replace the cervix quickly. My only thought is to get ice on my thigh. I throw a quick closure across the vulva using hog rings and small cotton umbilical tape. The hog rings only pinch a small piece of skin. They will tear out with a slight push from mamma.

"She should be able to tear this out when she calves, but you need to watch her closely," I instruct Duane as we hastily throw things back into the truck. I grab an ice pack out of the cooler and set it on my thigh as I start to pull out of the barnyard.

Spotting a cow out in the field with a pair of feet sticking out of the vulva, Joleen opens her window and hollers at Duane.

"How long has she been in labor?"

"Damnit, Joleen, I need to get this leg iced," I say with a frown.

"You can handle that, can't you?" I ask Duane. "She probably will pop that out with no problem."

"Oh sure, that is no problem for me," Duane says. "I didn't even know she was close."

My thigh has turned multiple shades of red by the time we get back to the office in Sweet Home. It is not the first time I've been kicked, and it probably won't be the last. It always seems that it is my left thigh. I'll limp for a few days with this one.

A Bear in the Backyard

Odie, our Chesapeake Bay retriever, stepped out the door onto the covered patio. His nose in the air, he sniffed the air. Making a muffled "woof," Odie knew the bear was there. He stood watchfully, waiting.

I noticed Odie's behavior. It was different from his usual bold bounce into the yard with a loud bark, announcing his dominance over his domain.

I looked, scanning the tree line of tall firs across the yard. Seeing nothing, I opened the door to speak with Odie. Just then, the hackles on his back stood up. I looked again, and there stood a large black bear at the edge of the trees.

"Odie!" I said. "Get in here."

The last thing I needed was to have Odie tangle with a bear.

Odie came back into the house, and from behind the closed patio door, barked loudly and jumped at the door, banging his nose on the glass.

"Aw, you're brave from this side of the glass," I said.

Odie wagged his tail. I swear he knows every word I say.

Unknown to us, the Oregon Department of Fish and Wildlife had captured a pair of problem bears in Corvallis area residents' yards. They had planned to transport them to the upper reaches of Quartzville Creek and release them there. Their plans were squashed with a heavy snowfall overnight. In their view, their only option was to release them onto some timberland on the eastern edge of Sweet Home. It only took a few days for our bear to establish his territory in our backyard.

Living near the top of a hill on the eastern edge of town, we were used the having a wide assortment of wildlife in our backyard. Deer were a constant fixture, along with raccoons, wild turkeys, and an occasional cougar passing through. But a resident bear was cause for both alarm and a lifestyle change.

Odie could cope fine. And he was a good signal as to how close the bear was to the yard. If Odie was cautious and quiet, the bear was close at hand. If he was loud and boisterous, the bear was off bothering one of the neighbors.

The cats were another story. We kept the cats indoors at all times. The exception was Charlie, our avid hunter, who insisted on being out most of the day and also most of the night. The other exception was the old feral tomcat who had adopted our backyard. He would have nothing to do with the house.

We were able to get along pretty well. The kids stayed inside most of the time. They only had to hear a few stories of people being mauled by a bear to convince them it was essential to give this bear a wide berth. I moved my rifle out from the gun safe and propped it in the corner by the patio door.

"I don't know what is more dangerous, the bear or your rifle propped in the corner," Sandy said.

"When I was growing up, there was a rifle and a shotgun in every house I knew," I said. "Kids knew it was not safe for them to touch a gun. So you just need to have a little lesson for the kids."

Then, one Saturday afternoon, the neighbor called.

"Dave, this is Herb. Can you come over for a minute? The bear is in my utility shed."

Herb lived next door in a modular home. He had a deck on one end of the house and small storage shed attached to the deck.

I went over to Herb's, avoiding the shed. I went to the front door. I had no more than arrived when the bear drags a 40-pound bag of dog food out of the shed.

Here is the bear, sitting on the steps leading up to the deck, legs crossed with the bag of dog food between his hind legs. He was scooping dog food out of the bag with his front paws and eating it like someone eating popcorn at the theater.

"What should we do?" Herb asked.

"In my opinion, my professional opinion, when a wild animal begins to display behavior that is unlike anything wild, it is time to shoot him," I said. "We can't have him rummaging around inside of sheds and garages. Next time it will be a house. Maybe we should call the state police."

Herb calls the state police.

"They say not to shoot him," Herb says. "They said to make a loud noise and scare him off."

"A loud noise, like what my rifle makes?" I asked.

"I have some firecrackers," Herb says. "An M80 should make a loud enough noise."

"That should be enough noise to make him jump," I said.

Herb retrieved an M80 and opened the back door. The bear was still sitting there, eating his dog food popcorn, without a care in the world, utterly oblivious of us. Herb lit the firecracker and tossed it toward the bear. It landed on the deck, only a few inches from the bear's butt.

"Crack!" The firecracker explodes. In a blink of the eye, that bear was 20 feet up a fir tree located 30 feet from the deck. Had he came in our direction, he would have been on top of us before we had a chance to move.

After this event, I called the state police game officer myself. I had spoken with him on other occasions.

"This is Dr. Larsen," I said. "I am one of the families up on Fiftieth Avenue who is dealing with this bear in our back yards. He is starting to rummage through sheds and garages. In my professional opinion, I think it is time to shoot him before someone walks in on him and gets mauled."

"I'm not going to give you permission to shoot him if that is what you are asking," the officer said.

"I'm not asking for permission. I am telling you what I am going to do," I said. "I'm okay with letting a judge decide how valid my professional opinion should be considered."

"Give me a day or two, and I'll have the Fish and Game guys get up there and recapture the bear," the officer said.

"I'll give those days unless I find him inside a building again," I said.

It took a few days, and they recaptured the bear. The second bear had been causing similar problems over on a hill on the other side of Wiley Creek. They captured him at the same time. I was told that because of the heavy snow in the high country, they released both bears out in the middle of the valley.

It was only a few weeks later when there was a story in the newspaper. Both bears had moved into Junction City and were causing havoc. The Fish and Game people had to shoot both bears.

All the Better to See You

RC was a big orange tabby cat. I had first treated RC for a severe fracture of a hind leg when he fell from a tree. Repair of that fracture required a pin and many wires, plus four weeks of cage rest. He was a friendly cat but displayed utter self-confidence when he was in the clinic.

He sat upon the exam table, watching for me to come through the door when I entered the room.

"My, what an ugly eye," I said. RC's right eyeball was swollen and bulging out from under his eyelids.

"I noticed this a little bit yesterday, and then this morning it was like this," Nancy said. "It is painful if I try to touch it, but he seems to tolerate it well when he is left alone."

"This is an advanced case of glaucoma," I said. "It is a little unusual to see this occur with no prior warning."

"Do we have any treatment options?" Nancy asked.

"We have a couple of options with an eye like this," I said. "We can go to Corvallis to see the veterinary ophthalmologist, or we can remove the eye."

"What is the ophthalmologist going to do?" Nancy asked.

"There are a couple of surgeries that can be done to save the eyeball, and maybe its vision," I said. "However, with an eye that looks like this, the vision is probably already lost. She can also offer evisceration and implantation of a silicone prosthesis instead of removal of the eyeball. Some people think that gives a better cosmetic appearance. A blank, nonfunctional eyeball remains."

"And what do we have when you remove the eye?" Nancy asked. "He will still have one eye, so I guess it won't change his vision much."

"I remove the eye and all the associated structures, including the eyelids' margin," I said. "When things are healed, we have a blank slate. If I do it right, that side of the face is smooth. If there is not enough dense tissue to close over the eye socket, there may be a little caved in appearance over the socket. Most cats get along fine with one eye."

"I think that we will just have you take the eye out here," Nancy said.

With that decided, we removed RC's right eye. Dr. Maxwell had recommended that I submit that eye for a pathologist to look at. She thought it was unusual for such a sudden onset of advanced glaucoma.

I had the results from the pathologist when Nancy returned with RC for suture removal.

"He is absolutely normal," Nancy said. "He does everything he did with two eyes. He still climbs his tree and everything."

"I am not sure that he has always climbed that tree too well," I said with a chuckle, remembering the fracture from years ago. "The pathology report says he had an autoimmune problem in his eye. They say there is a possibility that he may develop the same problem in his left eye."

"I guess we will cross that bridge when it happens," Nancy said.

***

In the following years, RC seemed to have more issues than I would expect to see in a middle-aged cat. But everything was manageable, and he was not bothered by the loss of his right eye.

It was almost three years to the day that Nancy rushed RC through the door. His remaining left eye had literally exploded overnight.

"He seemed fine last night," Nancy said as she caught her breath. "Then I looked at him early this morning, and I could see that his eye was getting big like his other eye. A few hours later, I look, and this is what we have."

"It looks like we have crossed that bridge you mentioned years ago," I said. "This eye is going to have to come out. There is no saving it now."

"Oh my! How will he get along, being totally blind?" Nancy asked.

"To be honest with you, I haven't had too many patients who were totally blind," I said. "I did have a client with a calf that was born blind. It had no functional eyes. It did just fine. It knew the pastures, knew where the feed rack was and where the water was. It could go in and out of the barn as long as it was with another animal or two. That calf grew up and had several calves, all born with normal vision."

"RC is so active, it will break his heart if he can't go out in the yard and climb his tree," Nancy said.

"I was at a veterinary conference several years ago," I said. "One of the speakers was a veterinary ophthalmologist. He told the story of his cat, who, it turned out, was totally blind. He said they had dinner guests at the house one night, and the guy noticed that the cat was blind. The doctor had no clue. That cat had lived in the house for several years, and the ophthalmologist had not noticed that it was blind."

"Okay, let's get it done," Nancy said.

The surgery was done, and RC went home with a blank slate for a face. It was a little eerie when he came back for suture removal. Sitting up on the exam table like he always did, he followed my every move with his 'blank slate.' Just like he was watching me.

"He is outside playing in the yard, just like he has always done," Nancy said. "Yesterday, he was even climbing his tree. We are so pleased that we didn't make the decision to put him to sleep."

RC lived an almost everyday life. As he aged, like many cats, he had his share of problems. Whenever RC was on the exam table, he 'watched' every movement I made. When he was in the clinic for hospital treatment, he would sit in his kennel and 'watch' everyone in the room.

RC lived a couple of months short of his eighteenth birthday and died of chronic kidney disease. Chronic kidney disease is the number one killer of cats over fourteen years of age. In those years, a male cat, neutered or not, was most unusual living to eighteen.

The Thomas Splint

The heat was stifling, and the room was packed. The air conditioner just couldn't keep pace.

"I hope he finishes this up a little early," I say to the guy sitting next to me.

He loosens his tie. "Yes, we all need to get out into some fresh air."

The speaker, a short, gray-haired orthopedic surgeon who teaches at Ohio State University Veterinary School, starts to field some questions from the audience.

"If nobody asks anything, we are out of here," I say to the guy next to me. He ignores the comment but unbuttons the top button on his shirt.

Then comes the first question, then another. "Didn't these guys listen to the lecture?" I say, more to myself than to the guy next to me.

"They must be his residents. They can't be wanting to stay in the sweatbox any longer," says my new friend in the tie.

Then comes another question, "What about using a Thomas splint on lower leg fractures?" some guy in the front row asks.

"I went to school over twenty years, and I never sat in the front row one day of all that time," I say.

The guy looks at me out of the corner of his eye but doesn't say anything.

"I haven't used a Thomas Splint in 25 years," the Professor says. "If you are going to repair a fracture, you should repair it the right way."

"He is a long way from the real world," I say to this guy next to me.

"What do you mean by that comment?" The guy says, almost like I said something that upset him.

"I mean, he would starve to death in Sweet Home," I said. "Everybody doesn't have three thousand dollars to go to a university for a fracture repair on their dog. What do you suppose happens to those dogs?"

"We can't take care of the world," this guy says, tightening his tie.

"We don't take care of half the dogs in this country," I said with a stern voice. "It is great to sit here and learn how to repair a fracture with equipment that only a fraction of the clinics in the state can afford. But when push comes to shove, you better be able to apply a Thomas splint. There will be a little girl who will be heartbroken if her dad puts her only friend in the world to sleep because he cannot afford a surgical repair."

"And what do you do when the repair fails?" the guy says as he slips back into his sports coat.

"You can say at least we gave it a shot," I said. "Then you better go back to school and learn how to do it correctly. I have used Thomas splints on everything from a six-week old kitten to a seven hundred-pound cow. I haven't had a failure. There have been a few legs that healed a little crooked but functional."

"Let's slip out of here, and I'll buy you a beer while you tell me a couple of cases," my new friend says.

"Okay, but you have to agree to one thing first," I said.

"And what do you want me to agree to?" he said.

"You have to take that damn tie off if you are going to buy me a beer," I said. "We had to wear a tie every day in vet school. I haven't worn one since. Probably won't until my daughters get married."

We were in the back of the room, so getting out the door without disrupting the class was easy. This guy takes his tie off as we head to a little bar in the hallway.

"This feels better," he says as we find a table.

I am not sure if he means having his tie off or if he is talking about the cool air in the bar. I finally notice his name tag. He is a speaker and a professor at the University of California at Davis Veterinary School.

"How long have you been at Davis?" I ask while we are waiting for a beer. "I knew a guy who did an internship there."

"I have only been there a few years," he said. "What did your friend think of the internship."

"He died," I said.

"Oh, I'm sorry," he said. "How did that happen?"

"He crashed a small private plane," I said. "He would have been better off just going to work."

"You have my interest in your comments on the Thomas splint," he said. "Convince me that you know what you are talking about."

"I'll compare two cases," I said. "They were separated by a few years but are good illustrations. Both tibial fractures that involved about half the length of the bone shattered in the middle half of the bone. One from a gunshot and the other, we did not know what happened."

"Did you repair both with a Thomas splint?" the Professor asked.

"The first case was a large malamute who belonged to a nurse," I said. "He was chasing the neighbor's cows, and the neighbor shot him. Shattered the middle half of the tibia. When I first saw him, I stabilized the fracture site with a pressure wrap and a Thomas splint."

"That was probably better initial care than many dogs get in a small clinic," the professor said. "Then what happened?"

"The dog was brought in by a friend," I said. "When the nurse finally got there, and we reviewed the films, she wanted a surgical repair. I said that this repair was way over my head. At that time in Oregon, we had limited options for a referral. There was a surgeon in Eugene, and we sent the case to him. This surgeon, who I knew, was amazed when this one hundred and forty pound dog with a shattered tibia walks into his clinic. He repairs it with a plate and bone grafts. They have all sorts of complications and follow-ups, but the bone did finally heal. I don't remember. Maybe I never really knew how much the bill totaled. I think she paid something like four thousand dollars. This was in the nineteen seventies. That was a whole lot of money."

"I have seen similar repairs," the professor said. "I would guess your estimate was close. And the problems with getting one of those fractures to heal are many."

"So, do you want to hear the other case?" I asked as I noticed that my beer was still mostly full.

"It must have been different," the professor said.

"The second case was a similar fracture in a blue heeler," I said. "We didn't know how the fracture occurred. The dog belonged to a girl who worked for me. This girl was a bright, good-looking girl who was in a poor marriage. She was in love with this dog. He was probably her closest friend in the world, and she had no money. She cried when we looked at the x-rays. The x-rays were probably identical to the first case, but no bullet."

"So you don't have many options at this point," the professor said.

"Very much between a rock and a hard place," I said. "I tell her, I cannot repair this surgically. She says there is no way she can pay for a referral; her husband would kill her, she says. I guess I believed that was probably more true than I wanted to know."

"So you put this leg in a Thomas splint," the professor said.

"We discussed options rationally," I said. "She was done crying. I said the best option of sending her for surgery was not an option, so what else can we do. Number one, we can cut the leg off. This girl was okay with that, having seen other three-legged dogs. But she wanted to hear the other options. Number two, we can put her to sleep. There were more tears now. She didn't want to talk about doing that. Then number three, we can put the leg in a splint. She thought there is no way this fracture was going heal in a splint. I teach my help well. So I say that when a splint works, it works well. If it doesn't work, which that is a possibility with a fracture like this, then we can fall back on the amputation."

"So you put the leg in a Thomas splint," the professor said again.

"Yes, I put the leg in a Thomas splint," I said. "I checked the leg weekly, only because she worked for me and it was easy to do. At eight weeks, the leg was healed. I left it in the splint for another two weeks, just for insurance. The leg was straight and functional, and the bone was thickened with a lot of callus formation.

The girl was ecstatic. I think I charged her only for the expendables that came to far less than a hundred dollars. The leg healed better, and in less time, than the leg on the malamute."

"Do you think we should put all these orthopedic surgeons out of business?" the professor asked.

"I never said that at all," I said. "But I think for him to stand up there and say that he hasn't used a Thomas splint in twenty-five years is condescending. The Thomas splint has fixed more fractures in years past than he ever will. And in veterinary medicine today, there is still a place for it."

"You make a good point," the professor said. "When we came in here, I thought I was going to have a lesson to teach you. I apologize. I think it was the other way around.

Link to Thomas splint: https://tinyurl.com/2zjdp5zh

The Turpentine Compress

I carefully started removing the wrap on the hind foot of Lady, a large golden retriever. This wrap was composed of a couple of rags and what looked like an old tee shirt.

I recoiled from the odor.

"When did this happen?" I asked Ralph.

"She got ran over yesterday afternoon. The foot looked so bad, and she was licking it so much, so I put a turpentine compress on it last night."

"Turpentine!" I said. "What made you think turpentine would be helpful?"

"My grandfather always used turpentine on a wound on his cattle," Ralph said with no hesitation.

I pulled the final pieces of cloth off the wound. Lady just laid her head down on the table, obviously resigned to the misery she was suffering.

This wound was a total loss of the skin from the hock down to the footpads. The skin was totally gone. The bare tissue looked almost cooked from the turpentine. I can only imagine how that must have felt when it was applied.

"Your grandfather maybe used it on some minor abrasions of the skin. That was maybe done in those days. Using it on a wound like this was nothing less than torture for Lady."

"Gee, I am sorry, Doc," Ralph said. "I was just trying to do something helpful."

"This wound was probably not repairable last night. It is definitely not repairable at this point. This leg is going to have to be amputated."

"When are you going to be able to do that surgery?" Ralph asked.

"We have a busy schedule, but I can't allow Lady to sit in a kennel and suffer waiting for surgery. We will move things around and get her into surgery right away."

"I don't know, Doc," Ralph said. "I am not sure she will be able to get along with only three legs. Maybe we should just put her to sleep."

"Dogs wake up in the morning and evaluate their situation. They just get up and go with what they have. They make do. People worry more about amputations than the dog. You will probably have to count legs to know that she only has three legs."

"What do you think it is going to cost?" Ralph asked.

"After what you put this dog through for the last twelve hours, you owe her whatever it costs. You owe her another few years."

"Doc, I was just asking to make sure I can pay the bill," Ralph said. "But if I decide to put her to sleep, that is the way it will be."

"Ralph, there is not a judge in the county who will doubt me if I say this is a case of animal abuse," I said. "Now, I don't believe in reporting any charges of abuse when it was done out of ignorance, and the person seeks care. That applies in this case. Especially since you are here this morning and not next week. But if you want to put her to sleep, my opinion might change."

"I am just trying to see where I stand," Ralph said. "You do the surgery. I will leave a deposit at the front desk. When do you think Lady will be able to go home?"

"Going home will depend on how well she is doing in the morning. If she is up and about, I will probably send her home. She will need a few pain pills, but she is probably going to feel so much better with this foot gone that she will be bouncing around."

"Where do you take the leg off, Doc?" Ralph asked.

"We could save you a little money by just taking it off at the hock joint. But Lady will do a whole lot better if we take it off in the middle of her thigh."

"Let's do what is best for her," Ralph said. "You have me feeling bad enough already."

We hurried Lady to surgery. The stench of the turpentine was nearly overwhelming in the surgery room. The surgical prep on the leg did little to lessen the odor or fumes.

I did a standard mid-femur amputation. This started with a bivalve incision through the skin to provide big enough flaps for closure over the stump. Extra skin can always be trimmed off the flaps as needed, but if the flaps were cut too short, it would be a significant problem to shorten the stump.

I isolated the major vessels first, double ligating the femoral artery to reduce any significant blood loss as the surgery progressed. After the vessels were ligated and severed, I separated the muscles quickly. I wanted to get this leg off and out of the surgery room as soon as possible, so everyone could breathe again.

When I finally had dissected down to the femur, I severed it with a Gigli wire saw. Once free from the body, I handed the leg off to be removed from the room.

At this point, I could take a deep breath and start closing the muscles over the bone to provide a well-padded stump. Then the skin was trimmed and closed. I had been a little faster than usual because I wanted that stinky foot out of the surgery room. It was a pretty brief surgery, less than an hour.

Lady recovered with some pain medication aid, but she had to feel so much better with that foot being gone. She was up and around by the early afternoon. Being a little older, it took her a few tries before Lady could handle getting up and down with only one hind leg. After those first few tries, she was acting like she didn't miss the leg much.

When I could feel confident that Lady would be ready to go home in the morning, I gave Ralph a call.

"Ralph, you can pick up Lady any time in the morning," I said. "She is feeling much better and is getting around on the one leg just fine."

"That sounds good, Doc," Ralph said. "I will be there first thing in the morning."

When morning came, Lady ate a good breakfast and walked on a leash well.

Ralph took care of the account and led Lady out to his pickup. He lowered the tailgate, and Lady started to jump into the back. She made a jump but did not have the strength in the one hind leg to make it up to the pickup bed. Ralph was quick to catch her before Lady fell. Then she jumped up with her front feet on the tailgate and looked to Ralph for a boost. One little boost from Ralph and Lady was in the bed.

"She will do just fine," I said to Sandy as I watched from the office. "Just give her a couple of weeks."

Lady continued to do well. As predicted, one had to count legs to make sure she only had three legs when she returned for suture removal a couple of weeks later.

After the sutures were removed, I never saw Lady or Ralph again. That was not a surprise. I had been stern with Ralph, and for him to find another clinic was almost expected. Perhaps it was from embarrassment for his actions, but more likely, it was from the veiled threat of reporting his abuse.

A Hasty Exam

"Dick, give me a quick rundown about what happened here," I said as I was unloading a few things from my truck.

Dick had called earlier, a little frantic. He had just pulled a calf from a cow and had a lot of problems with the pull. Now, he said, the cow had a prolapsed uterus.

"I have pulled a lot of calves, Doc," Dick said. "Now, don't get me wrong, I know you are supposed to be pretty good at the OB stuff, but I have done my share. This calf had both hind legs back, just his butt at the birth canal. You know, I called Dr. Jones from back in Montana, and he said to just push the butt forward and reach down and grab the legs. So that is what I did, and it wasn't as easy as he made it out to be. But I got the calf out, and it was dead. Then I looked at the cow, and her uterus is hanging out, just like you can see now."

"A lot of backward calves are dead when you get them out," I said. "Especially if they are breech like you describe this one. The calf just doesn't engage the cervix to cause the cow to start contracting. So we often just don't notice the cow is in labor until after the calf is dead."

"Well, now I am worried about the cow's prolapse," Dick said.

"What did Dr. Jones have to say about that?" I asked.

"He said I better give you a call," Dick said.

I started washing the cow up, and I tied her tail out of the way with a twine tied around her neck.

"This doesn't look good," I said. "This is just her cervix hanging out of her vagina. I am not sure I have ever seen this before."

I pulled on an OB sleeve and pushed the cervix back into the vagina. My hand just advanced through the cervix into an open space. There was no uterus attached to the cervix. I reached deeper. I could feel gut, rumen on the left, and even her right kidney. I reached down. There, in the bottom of her abdomen, was the uterus. I could grab it, but I could not pull it to the pelvis on the first try. Dick had ruptured the uterus in pushing the calf forward. It is sort of amazing that he could get the calf out of there.

"Dick, this cow is in bad shape," I said. "You tore the uterus completely off the cervix. It is lying in the bottom of her abdomen. She has got to be losing a lot of blood."

"What are we going to do now, Doc?" Dick asked.

I walked around to her head. Lifting her muzzle up, I looked at her oral membranes. They were ghost white. I kicked myself. I should have looked at her first before just concentrating on the obvious cervical prolapse.

"Does this chute have a side release, Dick?" I asked.

"Oh no, Doc, this chute is probably older than the two of us combined."

"Well, the first thing we are going to do is get this cow out of the chute before she drops dead," I said.

"This is a pretty big pasture. We might have a hard time catching her," Dick said.

"You're going to have on heck of a time getting her out of this old chute when she's dead," I said as I popped open the headgate.

The cow staggered out of the chute. She walked half a dozen steps, stopped and stood for a moment, and then fell over, dead when she hit the ground.

"What the hell happened?" Dick asked.

"Dick, when you were pushing that calf's butt forward, you pushed a little too hard," I said. "You tore her uterus right off the cervix. That pregnant uterus has some massive blood vessels, so she has been bleeding inside since that happened. Couple that with some shock, and you have a dead cow. There was probably no saving her once the rupture occurred."

"What would you have done differently, Doc?" Dick asked.

"Tissue feel is something that is learned," I said. "I probably would not have pushed his butt in the first place. Not that that was wrong, some vets do that. I am usually able to get the hind legs up without pushing the butt that much."

"Well, maybe Dr. Jones didn't explain things good enough," Dick said.

"You get what you pay for Dick," I said. "Had you called me first, the calf probably would have still been dead. But the cow would be alive. You need to decide who is going to do your veterinary work, me, or someone else around here, or Dr. Jones in Montana."

A Lesson Well Learned

Floyd was waiting for me at his driveway on Liberty Road. Sitting in his old pickup, he woke with a start when I pulled in alongside his truck. Floyd was a big man, his hands spoke of a lifetime of hard work, and his gray hair told of his age.

"I am glad you could come this morning, Doc," Floyd said. "I have a dead ewe in the upper pasture. She was fine last night."

"You saw her last night?" I asked.

"Yes, I'm not up there every day, but last night I was up there bringing a feed rack back to the shop with the tractor. I figured I wouldn't need it this spring and summer. The ewes were all there and their lambs. This morning, she is lying near the gate, all blown up like she has been dead for several days."

"That almost sounds like blackleg or something," I said. "I didn't think we were supposed to have that stuff on this side of the mountains."

"You can follow me up there," Floyd said. "The road gets a little rough, but it looks like your truck can probably handle it okay."

"Yes, you put four-wheel drive on these trucks, and they go just about anywhere," I said.

I followed Floyd up a well-graveled drive beside a cemetery. Floyd stopped and opened a gate to the dirt road to the upper pasture. I stopped and closed the gate as soon as I was through it.

"You can leave that open," Floyd yelled back to me.

"Habit, Floyd," I said. "My grandfather always said it was easier to close the gate than it is to wish you had closed it."

"Smart man, your Grandfather," Floyd said.

We continued up the badly rutted dirt road, finally crossing a cattle guard into the upper pasture. The ewe was right there beside the road. If Floyd had been up here last night, there is no way he would have missed seeing her.

This ewe was on her right side and blown up, not from rumen bloat, just blown up like a balloon. All four legs sticking straight out, there was some pinkish foam at her nostrils.

I got out of the truck and opened the vet box. After pulling on my coveralls and boots, I retrieved my necropsy knife, a bucket with some water, and a pair of gloves. As I approached the dead ewe, I noticed that she smelled like she had been dead for several days.

"You are sure she wasn't here last night?" I asked Floyd.

"She was sort of a pet, you know," Floyd said. "She was up butting me around when I was hooking up the feed rack. She was as normal as could be."

With gloves on, I pressed my fingers into her skin over her back. She had a lot of air under the skin, subcutaneous emphysema. It felt and sounded like I was popping bubbles. The wool pulled out along with the outer layer of the skin. There was a blueish discoloration to both the skin on both rear legs and extending along her back. This was blackleg.

"Floyd, I am not going to open her up at all," I said. "This is blackleg. It could be one of the other clostridial diseases like malignant edema, but that is sort of academic. They are all diseases that cause sudden death; most are impossible to effectively medicate, even if you make the diagnosis early. Most of the time, you just find them dead. The vaccines are all combination vaccines that include the whole bunch in one shot."

"I have not heard of blackleg around here," Floyd said.

"We saw a lot of it in Colorado," I said. "Guys had to vaccinate for it, or they would have dead animals. I never saw a case in Enumclaw. I was always told it was rare on this side of the mountains. This might change my thinking on that aspect of the disease. I am going to aspirate a little fluid from under the skin. I can pretty much confirm the diagnosis when I get back to the clinic and get the sample under a microscope."

"What should I do with her?" Floyd asked, pointing to the ewe.

"You should bury her deep if you have a backhoe or some grandkids," I said. "Bury her right where she lays, don't drag her anywhere. This disease is caused by spore-forming bacteria, and you would contaminate the pasture more. Or you could burn her if you have enough fuel. I don't know if the rendering company would pick her up or not."

"And what about the others?" Floyd asked.

"You need to get a vaccine into them," I said. "I have both a 4-way and a 7-way vaccine at the office. You should do that yesterday, and they will need a booster vaccine this year, in three or four weeks. I would hedge on the short side, three weeks."

"I can bury her with a backhoe," Floyd said. "And I can get some guys to help me vaccinate the others. That might take a day or two."

"I wouldn't let it take too long. You might find some more like this one," I said. "If you are going to work through the whole bunch, it might be a good time to worm them."

"You're trying to make a sheep rancher out of me," Floyd said.

***

Back at the clinic, I did a quick gram stain on a slide I made from the aspirate from the ewe. The slide was covered with large gram-positive rods that were characteristic of clostridium organisms.

That, coupled with the clinical signs of the ewe, was pretty much diagnostic for blackleg.

I checked and made sure I had an adequate supply of vaccine for Floyd.

The following day I thought I should call a couple of other veterinarians in Lebanon and Albany. I wanted to check to see if they saw any of the clostridial diseases in this area. As I was considering those calls, the telephone rings.

"Doc, this is Walt up on Fern Ridge," he said. "I went out this morning to look over the animals in the field. I check them morning and night. Everything was fine last night. This morning there are seven dead steers in the pasture. I sure hope you can tell what the heck is going on with that."

"Give me a little time, Walt," I said. "I can get up there in a little over an hour. Are they near the house?"

"No, they are down the road a bit," Walt said. "I will be in my blue truck and will park at the gate. You should not have any problems. The field is right along the road."

When I pulled up to the gate, I could see the seven dead steers scattered across the field. These steers were all at market weight—what a loss.

"I moved all the other stock out of this pasture," Walt said. "What do you think is going on?"

"Let me get a look at them," I said. "You said they were fine last night?"

"I was up here before dark last night," Walt said. "They were all up and running around like nothing was wrong."

All of these steers looked just like Floyd's ewe. Laid out on their sides, puffed up like a balloon, and all four legs were sticking out, straight and stiff. When I pushed on the skin, I could feel the air popping under the skin, like pressing on some bubble wrap.

The carcasses were in an accelerated rate of decay, with a strong odor, and hair easily slipped from the skin, especially in areas where the skin was discolored. These steers all died from blackleg. I took a few aspirates and went to talk with Walt.

"Walt, this is blackleg," I said. "Everything fits the diagnosis, sudden death, rapid decay, subcutaneous emphysema, and discoloration of the skin and tissues. If you want a thorough diagnosis, you need to load one of these guys up and take him over the diagnostic lab at Oregon State."

"What is the risk to the rest of the herd?" Walt asked.

"Clostridial organisms are spore-forming bacteria," I explained. "Those spores last for many years. They have spores from soil collected in Kansas in the eighteen-nineties that are still infective. What causes an event like this is not really known. What is known is that vaccination is a very effective preventative. You need to vaccinate the rest of the herd yesterday."

"Okay, I can get a crew together and do that chore this afternoon," Walt said.

"If you have any animals that have not been vaccinated before, you need to booster the vaccine in three weeks," I said.

"What do I do with these steers?" Walt asked.

"The rendering company can probably pick them up," I said. "If not, you should bury them deep, right where they lay, no dragging them across the pasture."

"Have you ever seen anything like this before, Doc?" Walt asked.

"In Colorado, the clostridial diseases were common, and ranches who did not keep up on their vaccines always had problems," I said. "There, I would see several animals at a time. In Enumclaw, I never saw a case. Had you talked to me last week, I would have recommended vaccinations for blackleg, but I would not have worried too much if you failed to take my advice. I looked at a ewe yesterday. It looked just like these steers. I think vaccinations will be required in the herds I take care of regularly."

"That sounds like you are a little strict, Doc," Walt said.

"Better than checking seven dead steers in the pasture," I said. "There are other veterinarians in the valley. People always have choices."

A Surprise Visit

"Dave, I am flying from San Francisco to Seattle this weekend," Marsden said on the phone. "If it works out for you, I could stop off in Eugene, spend the weekend and then catch a flight out of Portland."

"What a surprise," I said. "That will work great. I can pick you up in Eugene, and we can drop you off at the Portland Airport. We can take the back road to Portland, so you can see a little of western Oregon."

Marsden had been in the Army with me. We had both been in Company D at Fort Devens, a duty company, while we were waiting for class spaces to become available.

After Fort Devens, I went to Korea for thirteen months and then went to West Germany. Marsden was at Wobeck, a small outpost out of the village of Schöningen when I arrived. We worked together in the maintenance shop for the seventeen months that I was there. We worked well together, and along with a few others, we became quite a team.

In the early summer, Marsden stepped out onto the ramp of the small commuter airplane at the Eugene airport. My seven-year-old daughter, Amy, had made the trip with me to pick Marsden up. Her hair was whipping around in the stiff breeze.

"I bet you had a bumpy ride in that little puddle jumper," I said as I extended my hand. Marsden had not changed any in the ten years since we had seen each other. Still tall and slim with thinning rusty red hair. At least taller than me, but that is pretty easy to accomplish.

"It wasn't bad until we came down to approach the airport," Marsden said.

"I was hoping I could identify you," I said. "But you haven't changed a bit."

We had both returned to school after the Army. I became a veterinarian, and Marsden got a Ph.D. in geography. He was working for a German company now and living some in the US and some in Germany. His wife, Elke, was from Schöningen.

"How long do you have to stay," I asked.

"Not long, I have a flight out of Portland on Sunday evening," Marsden said. "I hope that doesn't inconvenience you much."

"No problem at all," I said. "We can find something to do on Saturday and take the back road to Portland on Sunday to give you a flavor of Oregon."

The only complication we had was my folks were staying with us this weekend. That made sleeping arrangements a little tight. We put my folks in Brenda's bedroom and Marsden in Derek's bedroom. Kids are adaptable enough that they can sleep anywhere.

Finding something to do on Saturday in a veterinarian's house is not difficult. The phone rang early. I had a cow to look at in Cascadia. She was down after calving the day before. She was out in the brush, so we would have to walk a bit. What an opportunity to show Marsden a slice of Oregon that few visitors get a chance to see.

"The only problem," Marsden said. "I don't have any old clothes to wear. I probably don't want to wear a suit out there."

"I think you will sort of fit some of my stuff," I said. "What size of shoes do you wear?"

"I wear a size 10," Marsden said. "It looks like I could probably squeeze into an old pair of your shoes."

So I decked Marsden out in an outfit that would fit right in at Cascadia. Old pair of Levi's, they could have used another three inches on the legs, but at least his pants legs wouldn't get wet. I found a shirt that fit and a wool shirt to keep him warm. We would be fine if the weather held.

The drive up the river was a pleasant one. In the early summer, the streamflow was still vigorous, and everything was green. Marsden was quiet. I was hoping he wasn't regretting his decision.

"I am impressed," Marsden said. "Your life is just what you said you wanted when you left Germany."

I hadn't thought much about that, but I guess he was right.

We pulled onto the place at Cascadia and were greeted by an old gray-haired lady and her son. The cow was down in a thicket of brush. We gathered my stuff and followed the son out to the cow. Turned out to be a milk fever. I gave her some IV calcium and a couple of other shots, and we were done.

"Since we are halfway to Mountain House, I will take you up and show you a big tree."

We continued our drive up the river and turned up Soda Fork Creek, right before we got to Mountain House. A couple miles up the creek, I pulled over, and we got out to look at a massive old-growth Douglas fir. This solitary tree was located between the road and the creek. I am not sure of its height, but it was ten to twelve feet in diameter at the butt.

"Are there any others like this one?" Marsden asked.

"Over the hill, on the Middle Santiam, there is a whole hillside."

"Can we go there?" Marsden asked.

"Sort of, to get into the trees, we would have to cross a massive slide," I said. "We are not really dressed for that trip. But we can get a look at the hillside. There is a big struggle going on right now, trying to make that entire area a wilderness area."

"I would like to go look," Marsden said.

"It is a little bit of a drive. We should go down to Mountain House and get some gas first. I could probably make it on what we have, but it is just a good idea to drive with a full tank in the backcountry."

We drove back down the road to Mountain House and entered the rustic combination, store, restaurant, and rest stop. They did have a gas pump.

"We need some gas," I said to the unshaven guy behind the counter.

"I am out of gas. If you come back on Tuesday, I can sell you all the gas you need," the guy said.

Marsden chuckled at that.

"We won't be around on Tuesday," I said.

"You might be able to get some at old man Totman's in Cascadia," the guy said.

We drove back to Cascadia. The old man was in the cluttered station. I motioned Marsden to look at the stuffed bobcat on the high shelf. It took some idle conversation and a couple of stories, but we got our gas tank full and headed back to Soda Fork.

The road over the ridge to the Middle Santiam River followed Soda Fork for several miles and then climbed the hill to Cool Camp. Cool Camp was probably once a logging camp location, but now it was just an intersection of logging roads.

From Cool Camp, it was all downhill to the banks of the Middle Santiam River. The road passed through timber and areas of regrowth on the harvested ground. There were many twists and turns in this road, so the five or six miles seemed much longer.

We came to a stop at the large slide. I had hoped that Marsden would have the panorama of the Middle Santiam's old-growth forest in full view. Unfortunately, the drive was somewhat in vain as the forested hills were shrouded in fog.

We spent some time here and then returned home for the remainder of the day: dinner that evening and some idle conversation over a beer. Sunday morning, we took Oregon Highway 213 to Oregon City. This provided a much better slice of Oregon than the drive along the freeway. I was trying to show the best of western Oregon, and Marsden was most impressed with the old car bodies in the back yards along the way. Those don't exist in the east, where salt is used on the highways in the winter.

From Oregon City, it was a short trip to drop Marsden off at the Portland airport. A handshake and we parted ways again, almost the same as in Germany, but this time it was Marsden catching the plane.

For Marsden's information, the area we spent those few minutes on the Middle Santiam is now located well within the Middle Santiam Wilderness Area's boundaries.

Agroceryosis, The Lack of Groceries

"What do you think, Doc," Al asked as we stood at the fence watching a skinny cow in the corral.

"It doesn't look like she has any diarrhea. When did you deworm last?"

"I gave her some pills I got at the feed store a couple of weeks ago," Al said.

This was my first visit to Al's place. Al in his fifties and dressed like he just stepped out of his office. My guess was he is a hobby farmer, probably a retired police officer out of California. I was a little surprised he had wormed the cow.

"Is she nursing a calf?" I asked.

"Yes, and the calf isn't doing really well either," Al said.

"Let's get her in a chute, and I will get a look at her," I said.

"A chute, what do you mean by a chute?" Al asked.

This was going to be a bigger problem than I expected. This guy doesn't know anything."

"I guess I better grab my rope," I said. "Are these posts strong enough to hold her?"

"I think so. They were here when I bought the place last summer. I bought this cow then also. She looked good when I bought her. She calved with no problem. Now she is skin and bones."

I threw the lasso over her horns and took a couple of wraps on the corner post. She wasn't wild, and I quickly pulled her up to the post and tied her close.

"What are you feeding her, Al," I asked?

"My pasture is dry as a bone now, but she gets all the hay she can eat."

I examined the old cow. She looked fine, except she was skin and bones. Her udder was almost empty, but the milk I stripped out looked fine. I collected blood and fecal samples.

"Al, what you call hay, in the feed rack there, is straw," I said as I turned the cow loose and crawled back across the fence. "I will look at these samples when I get back to the office. But her problem is agroceryosis."

"Agroceryosis, I have never heard of that. It sounds serious," Al said.

"It is serious, Al. If it is not corrected, it will surely kill her," I said. "What it means is there is a lack of groceries. That straw you are feeding doesn't have much in it. We need to have a little discussion on basic nutrition."

"That's why I called you. I need to learn all I can," Al said.

"How many cows do you have," I asked?

"She is my whole herd, her and her calf," Al said. "I figured I needed to learn with small numbers before getting a bunch."

"That was probably the best decision you could have made. We have a couple of months to get her ready for winter. Otherwise, I would have been out here on an emergency call when she was down and dying at the first snowfall."

"You sound pretty sure that is what would happen to her," Al said.

"Every year that I have been here, the first snow brings a rash of calls. Usually, it is a horse and a couple of cows. They all look the same. And they have inadequate shelter, no fat on their bones, not a snowball's chance in hell of surviving the night. The thing I have never been able to figure out is how the owners can figure out they are going to die tonight but can't figure out they need to feed them a little."

"Well, I could at least recognize that she needs something," Al said.

"Without getting out the nutrition books, you can probably understand that the cow has basic needs that need to be met. She needs protein, and she needs an energy source. Sometimes you can meet all those needs in good grass hay. That hay might contain eight percent protein. This straw probably has less than two percent protein. She can't eat enough to meet her protein requirement. They say she is bulk limited. It is sort of like you would be trying to live on lettuce."

"So I need to go shopping for hay," Al said.

"This cow is going to need more than just grass hay. She has to make up some ground before winter. And her milk production requires a lot of energy. That is one reason she is so thin. She has put all of her fat reserves into producing milk for her calf."

"So I need some grain also," Al said.

"Yes, you probably need some good grass hay, maybe a bale or two of alfalfa and a couple of bags of grain," I said.

"You think that will do it," Al asked?

"That and a mineral block. Then you have to change her ration slowly. You can change to the grass hay with no problem. But you need to get her on that first, then add a small amount of alfalfa and a little grain."

What do you mean by a small amount and a little grain," Al asked?

"You get her on good grass hay for a week, then start giving her a couple of cups of grain once a day. The third week, give her half of a flake of alfalfa on top of her grass hay. I will drop by sometime during that third week and just eyeball her. We should be able to see a change by then."

"What about those samples," Al asked?

"I will give you a call tomorrow. If you wormed her, I don't think we will see much in the samples."

"Okay, I will expect to see you in a few weeks," Al said.

"Does she have access to the barn?" I asked.

"Yes, but she doesn't seem to want to use it much."

"She will when winter gets here. You can take this straw and use it to bed down a stall space for her and her calf. You might be surprised at how she reacts when she has a bedded stall."

***

Several weeks later, I pulled onto Al's place as I returned from a call out at Crawfordsville. I waved at Al as he was pulling on his boots on the porch. I stopped out by the barn.

"Doc, she is a completely different cow," Al said. "I am embarrassed now that I almost starved her to death."

The cow and her calf were bedded down in the straw in the barn. They were both chewing their cuds and paid no attention to us.

"The calf never used to eat when I was feeding them straw, but now he bellies up to the feed rack and fights for position with his mother," Al said.

I could still see ribs on the cow, but they were covered with a layer of fat already. Her hip bones looked smoother now also.

"It looks to me like she will be fine. I would start giving her a full flake of alfalfa and give her some grain twice a day. When she starts to look like your neighbor's cows, you can slow down on the alfalfa and grain."

"You think she will be okay for this winter?" Al asked.

"I think she is going to be okay. Before you go out and buy the rest of the herd, you need to drop by the office, and we can go over nutrition in a little more detail. And we can discuss a vaccination and parasite control program that will work well for a small herd."

"Isn't it funny, I remember cows on my grandfather's farm, and I had no idea what kind of work and knowledge went into raising them."

"Those grandfathers didn't need a book. They just knew the cows. They made it look easy," I said.

"Thanks again, Doc. And I will remember to drop by the clinic before I get the rest of my herd."

Back to Her Old Self

"I know you don't believe me, Doctor," Althea said. "But she is just not herself."

Penny was a yellow Lab cross, maybe more reddish than yellow and smaller than most Labs. Her toenails were long, causing her to be unsteady as she thrashed about on the exam table. Penny was one of those dangerous, friendly dogs. If you got too close, she would lick you to death.

Althea was a slender middle-aged lady who noticed our clinic while she was doing laundry next door. At one time in her life, she had been a medical technician. Now she worked nights at the hospital, transcribing records.

"I am finding it hard to think she is sick," I said. "I don't see too many sick dogs dance around on the exam table like this. Let's draw some blood, and I will give her an antibiotic injection. We won't have the blood results until morning. You can take her home, and I will give you a call when I get the results."

"I feel a lot better with you running some blood on her," Althea said. "And I am sure she will be more comfortable at home."

"We can set up an appointment to recheck Penny in the morning, about nine. We should have the blood results by then."

Collecting a couple of tubes of blood from Penny was somewhat tricky. She was bouncing so much it was almost impossible to restrain her enough to get the needle in a vein. The blood collection was followed by an injection of ampicillin. Then I lifted Penny down to the floor. She was instantly dragging Althea toward the outside door,

"I will get the appointment written down," I said to a struggling Althea who was having trouble bringing Penny to a stop at the front desk. "We will see you both in the morning."

***

I scanned the values quickly when Judy handed me the results of the blood work the following day.

The white blood cell count jumped out at me, twenty-nine thousand white blood cells. There are not too many things that will cause that high of a number in the dog. But Penny was an older, intact female. A pyometra, a pus-filled uterus, was high on the list of possibilities.

"I think you should clear the schedule for the morning," I said to Judy. "I am betting that we are going to have to schedule surgery for Penny."

Right at nine, Althea came through the door with Penny. Penny was still bouncing around on the end of her leash.

"I know she still looks fine to you," Althea said. "But she is even slower this morning. And I noticed a little vaginal discharge this morning."

"Well, her blood results got my attention this morning," I said. "Her white blood cell count was twenty-nine thousand. In an older, intact female dog with vaginal discharge, the diagnosis is a pyometra, a pus-filled uterus, until I prove otherwise. We can easily confirm that diagnosis with an x-ray."

"I am on somewhat of a budget here," Althea said. "What are the treatment options, and can we do those without the x-rays?"

"There are no options. The treatment is surgery," I said. "We have to get that uterus out of there before we get into all sorts of complications. Surgery is diagnostic also. If we are correct, we save the cost of the x-rays. If it is something else, then we are back to square one. But doing an ovariohysterectomy while we are there would be to Penny's benefit."

"I think with my funds, we should just go right to surgery," Althea said. "You can call it an exploratory."

"That sounds good to me," I said. "I am pretty confident of the diagnosis, especially with the vaginal discharge this morning. We will give her a bottle of fluids before surgery and continue the fluids during surgery. It would be a good idea to keep her overnight, but we can make that decision this afternoon."

With that, we started getting Penny ready for surgery. Althea was right. She was slower this morning. As we worked through the preoperative exam, I noticed that the vaginal discharge was increasing. She was dripping a foul-smelling vaginal discharge onto the exam table.

We got an intravenous catheter into her cephalic vein on a front leg and started a bottle of Ringer lactate. I added a dose of IV antibiotics to the fluids, and we made Penny comfortable in a kennel. At the same time, we set up the surgery room. I wanted to get the whole bottle of fluids into her before we started surgery.

With the catheter and the fluids already set up, anesthesia was easily induced. A good dose of Pentathol, and then with an endotracheal tube, we hooked her up to the Metaphane gas machine. When we got her on her back, it was evident that her abdomen was full, and the uterine discharge was really going now.

Often a pyometra will start with a closed cervix. The pus that develops due to chronic overstimulation of the uterine lining from estrogen accumulates in the uterus. The uterus can reach a large size, at least as large as a full-term pregnancy, if not larger. When the cervix opens, a lot of the pus is discharged, and the dog may feel better for a time.

When we had Penny's belly prepped for surgery and a second bottle of fluids started at a slow drip, I pulled on a pair of gloves and opened the surgery pack. In the mid-nineteen-seventies, I wore a surgical mask, but I did not gown for surgery.

I started with a small midline incision. As soon as I could see the size of the uterus, I extended the incision to about six inches. Then very gently, I eased the right uterine horn out of the abdomen and laid it out on the surgery drape. Then I did the same with the left uterine horn.

This was one large uterus. In school, during my junior year in my surgery rotation, I had assisted with the ovariohysterectomy of a very pregnant Saint Bernard. With a third trimester pregnancy, that uterus was large. This uterus, on a much smaller dog, was just about that size.

I clamped the ovarian vessels, ligated them with 00 Dexon, used scissors to sever the broad ligament from the uterus along each uterine horn. There were a couple of larger vessels in the broad ligament that needed ligation. Then I clamped the uterine body at the level of the cervix and ligated the middle uterine vessels. I separated the uterus between two clamps. With great care, not wanting to rupture the uterus at this point, I transferred the entire uterus to a disposal bucket. I took a deep breath with that container of pus safely disposed of in the bucket.

Now it was simple, I oversewed the uterine stump and returned it to the abdomen. Then, to make sure there was no contamination from the uterus, I changed gloves, surgery pack, and drape. Then I closed the abdomen in a standard three-layer closure.

With the slow recovery from metaphane, I turned off the gas when I started to close the abdomen. By the time I finished closing, and we got Penny cleaned up and back to the kennel, she was just beginning to wake up, and I was able to pull the endotracheal tube.

Wow, just wow. Penny wakes up, and she is back to her old self. It is always amazing when you remove a bucket of pus from a dog's abdomen. Then give her some antibiotics and a couple of bottles of fluids. With Penny, she was bouncing off the walls of the kennel.

We pulled the catheter and checked her over. Everything was in order, and the incision looked good.

"You need to give Althea a call," I said to Judy. "I think we can send Penny home anytime this afternoon. She will surely be quieter at home than she is here. We will check her in a day or two, depending on Althea's schedule, to make sure things are going along okay."

Althea was pleased. Penny was jumping all over herself when Althea showed up. I was happy to see Penny pull Althea out the door.

The recheck was quick. Penny was almost uncontrollable. I mainly wanted to check the incision. It was fine.

"I told you the first time you saw her," Althea said. "Now, you can see that I was comparing her to her regular activity level."

Benefits of Experience

The phone jarred me awake. It was going to be another late night, and I thought calving season was over. I glanced at the clock as I picked up the receiver, one-thirty.

"Doc, this is Jack. I have a llama who gave birth tonight," Jack said in an excited voice. "They almost always birth during the morning, but this one came tonight, and she has a problem."

"What's going on with her, Jack?" I asked.

"I think she has a prolapsed uterus," Jack said. "And she is not doing well. She hasn't even looked at the baby."

"Llamas rarely have birthing problems," I said. "But when they have problems, it is when they give birth at night."

"Can you come out and get a look at her, Doc?" Jack asked.

"It won't take me too long," I said. "You know I am in bed."

"I know, Doc," Jack said. "I'm sorry, but it is like when we used to have to Hoot Owl in the woods. You just got to get up and go."

"Don't do anything with her until I get there," I said. "If she is not feeling well, she might be in shock. We don't want to add any stress. Where do you have her, Jack?"

"She is in the little barn up here by the house," Jack said. "I will have all the lights on for you."

I have seen very few birthing issues with llamas. One with a prolapsed vagina that I ended up delivering the baby, and that was about it. If this is a prolapsed uterus, it will be a first for me. Unlike a cow, this llama is probably worth about thirty-thousand dollars.

Jack was waiting at the barn door when I arrived. He was having trouble standing still.

"I am glad you could come so quick, Doc," Jack said as I stepped out of the truck. "She doesn't look good to me at all."

I gathered my stuff for the first trip into the barn: a stethoscope, bucket of warm water, Betadine scrub, and a dose of Oxytocin. Jack had the cria under a heat lamp. Mom was paying no attention to the baby. That in itself was an unfavorable sign.

She did have a prolapsed uterus. The membranes had already passed. Just one horn of the uterus was prolapsed. But Momma did not look good. Her oral membranes were ghost white, and she was resting on her sternum and was not responsive to my attention.

"Jack, there was a day that I would blame her condition on blood loss, but I think she is in shock," I said. "Let me run back to the truck, and I will get a couple of bags of fluid and some medication, and we will see if that helps her before we do anything with this prolapse."

"Is she going to be able to breed after this?" Jack asked.

"We have to worry about her surviving the night before we worry about her breeding again," I said. "Actually, Jack, I doubt if there is much data on fertility in the llama following a prolapse. Most of the time, they breed back in the first month or so following delivery. I would think that is not going to happen, but I have no data or experience on the top of my head to support any opinion."

I hung a bag of fluids from a nail on a nearby support post for the barn and placed a fourteen gauge needle in her jugular vein. I ran the fluids as fast as they would flow and added twenty milligrams of dexamethasone sodium phosphate to the first liter.

Halfway through the second liter, mom was looking for her cria and acting like she would live. I slowed the flow and gave ten mg of oxytocin IV. Then I turned my attention to the prolapse.

This was a fraction of the size of a bovine prolapse, and the oxytocin was already contracting the uterus. I scrubbed it vigorously with Betadine Scrub. Then I lubed it with J-Lube, a powdered lube that, when wet, became as slick as anything I knew.

After a couple of pushes the uterus popped back into its normal position. I ran my hand through the cervix and made sure the uterine horns were returned to normal position. At the same time, I put a couple of grams of oxytetracycline powder in the uterus. Then I sutured the vulva closed with several sutures.

I have never experienced a prolapse that came out a second time. Especially if oxytocin was given to contract the uterus. But, be it training or just making myself and the owner happy, I always sutured the vulva for one to three days.

I cleaned up mom and removed the IV. I gave a good dose of long-acting antibiotics, and she jumped right up, looking for her baby.

"My guess is we are home free, Jack," I said. "I'll run by and recheck her in a couple of days and get those sutures out. You just need to check her over real well in the morning. Make sure she is eating and taking care of the baby. You call if you have any questions about how she is doing."

The trip home at three in the morning gave me time to ponder. Would I have been so quick to provide fluids to this gal if I had not been able to peek inside the belly of Ag's cow a couple of summers before? I am not sure of the answer to that question. I guess once you know the correct answer, it is hard to think of another solution.

Rosebud's Wire

Hardware disease, which results from the indiscriminate eating habits of the cow, was frequently seen when hay was baled with wire. Any stray sharp metal, like a nail, could be involved. The hardware will fall into the reticulum, considered the second stomach. The cow's heart lies just on the other side of the diaphragm from the reticulum. Then when the wire pierced the reticulum and the diaphragm, it would poke into the heart. This can cause acute discomfort and rarely rapid death, but more likely a slow death from an infection around the heart and chronic heart failure.

My first recollection of a cow with hardware was with our Linda cow at Broadbent. Named after my sister, Linda cow was a Jersey cross cow. Of all our cows, she was the least attractive. Probably crossed with a Guernsey, she was white with some orangish brindle coloration.

In the early nineteen-fifties, hardware disease was treated with surgery. The rumen was opened on the left side. The operator would remove some of the content and then reach anteriorly to the reticulum and extract the offending wire. Dr. Crawford, a veterinarian from Coquille, did the surgery on Linda. She carried an ugly scar for the rest of her life, which was not unusual with that surgery.

By the nineteen-sixties, hardware disease was treated with a magnet given orally to the cow. The magnet would fall into the reticulum, secure the offending hardware, and withdraw it into the stomach through regular stomach activity. It would stay there, with the magnet until it rusted away.

One valuable cow was treated at Colorado State University Veterinary Hospital while I was in school. It was suffering from severe pericarditis (infection around the heart) from hardware disease. In a last-ditch effort to save the cow, they removed one rib section and opened the pericardium to the outside. They could flush this area a few times a day. The problem was a large amount of inflammatory tissue in the pericardium, constricted as it aged, and the treatment failed.

My first case of hardware disease after graduation occurred in Enumclaw. Andy had called with a steer that was not doing well.

"I don't know what is going on with him, Doc," Andy said. "He just stands mostly and walks sort of stiff-legged when he does move."

Andy had the steer in his old dairy barn, just loose in the stanchion area.

"Let me get my rope, and we will get a look at him," I said.

When the lasso landed over his head on the first throw, it surprised both the steer and me. He jumped to the right, and then back to the left, and then he fell over dead.

"What the heck happened to him?" Andy said in amazement.

A quick necropsy showed a five-inch wire poking into his heart. Had I been able to restrain him quickly, a magnet possibly could have saved him.

"My guess is you are feeding hay baled with wire," I said.

"Yes, I just got a new load of alfalfa," Andy said.

"I would suggest you give all your other cows a magnet," I said. "It is pretty easy to do, and it is a good preventative for this type of thing. Although it is rare to see a steer drop dead like this one."

"So, was that in the hay, or did I do that when I opened the bale," Andy asked.

"There is no answer to that question," I said. "I guess it doesn't matter to this guy."

"Can we eat him?" Andy asked.

"I guess that depends on how hungry you are," I said. "He doesn't qualify according to the food inspection criteria. He probably had a temperature and some inflammation. Obviously, you could eat him, but I don't think the meat will be very good."

***

Barney was my very first call in Sweet Home. I had only been in town for a couple of days when the phone rang.

"Doc, this is Barney. Stan over at the feed store told me you were in town. I have a cow that stopped eating a couple of days ago, and she is just not doing very well. I sure would like for you to get a look at her."

"Barney, I am really not ready to do any calls," I said. "I don't have most of my equipment and drugs yet."

"If you could just look at her," Barney said. "If I have to call Albany, it will take them three days to get out here. It is going to be great, having you in town."

"Okay, I will get a look at her, but no promises," I said.

Barney's place was up Ames Creek. He had several acres and a couple of cows. I pulled into the driveway in our car as I didn't have a truck yet. Barney was waiting to greet me.

"Hi, I am Barney. I hated to pressure you, Doc," Barney said as he shook my hand. "But you have to realize how difficult it has been to get a veterinarian out here in the past. Everybody is excited that you are coming to town."

"I'm glad to meet you, Barney," I said. "I'm here. It's just that I'm not quite organized yet. Shipments of equipment and supplies are coming every day, but the clinic is way behind schedule. The little two-bedroom apartment we have rented is bursting at the seams. We have boxes stacked to the ceiling. But let's get a look at this cow of yours. Oh, by the way, Barney, you are my very first client in Sweet Home."

Barney's cow, Rosebud, was a Hereford. She was in good shape, in fact, probably a little overfed. I would guess she was five or six years old. Doing an exam showed a slightly elevated temperature and much reduced rumen motility. Everything else was within normal limits.

"How long has she been sick, Barney?" I asked.

"I noticed that she stopped eating a couple of days ago. She has just been standing around, not moving much."

I squatted beside Rosebud on her left side. I leaned against her belly with my elbows on my knees. I placed both fists in the center of her stomach, close to the edge of her ribs. Then with a hard push upward, like a punch in the gut, I jiggled her entire abdomen.

Rosebud let out a noticeable groan. Almost a diagnostic pain response for hardware disease.

"She has a wire, Barney," I said.

"A wire, what are you talking about?" Barney said.

"Cows aren't very discriminating when they are chewing on hay," I said. "If there is a wire in that hay, it goes right on down. It falls into the reticulum, a little pouch on the front of the rumen. If the wire happens to puncture the wall, it is a very short distance to the heart."

"So, what do we do about it?" Barney asked. "It seems I have heard about them doing surgery because of a wire."

"At this stage in the game, we rarely have to do surgery," I said. "I will give her a magnet and some long-acting sulfa boluses. I just happen to have both of those. That should take care of things in a day or two."

"That sounds better than surgery," Barney said. "Do I have to do anything with her?"

"You just keep an eye on her," I said. "Call if she is worse tomorrow. I will give you a call or probably just drop by when you get off work in a couple of days. Since you are my first call, I don't have a lot to do just yet. I should have plenty of time to look after her."

***

Two days later, when I stopped at Barney's, Rosebud was out in the pasture grazing. Barney came out of the house, all smiles.

"Doc, Rosebud is acting like nothing was ever wrong with her," Barney said. "I can't thank you enough, and I am so glad that you came to town. We will see you again, I'm sure."

"Thanks, Barney. Just give me a call when you need me."

Choose Your Surgeon with Care

Marilyn carefully placed Dorie on the exam table. Dorie, a pretty, tabby and white female kitten stood, unmoving, on the exam table. Her stance was rigid, not what I would expect to see in an eight-week old kitten.

"She was fine yesterday when I brought her home," Marilyn said. "I noticed she was a little reluctant to move last night, and she was a little painful when I handled her. But this morning, she will hardly move, and she cries out at the slightest touch."

"Where did she come from?" I asked.

"I adopted her from a humane society over on the coast," Marilyn said. "They insisted that they spay her before they would let her out the door. I pleaded with them to allow me to bring her to you, but they would not let her go."

"She is pretty young for a spay," I said. "But it is becoming a thing with the humane groups. They struggle with people not getting the surgery done."

"I understand that," Marilyn said. "But she is so tiny, waiting a few weeks would have been better. I tried to get them to call you as a reference, but they were adamant, they would not let her out the door without the surgery."

I picked Dorie up as carefully as I could. She cried out with a soft cry. The incision looked fine. It was a very short incision, only a little over a centimeter, which means the surgery was done with a spay hook. Most veterinarians use a spay hook during a spay. It is a blunt hook instrument to retrieve the uterus from the abdomen. I virtually never used one. I always used a finger, something I learned from Dr. Ferguson, a surgical resident at CSU when I was there. It required a slightly longer incision, but that was of little consequence. Incisions heal from side to side, not end to end. I always felt more comfortable with a feeling finger retrieving the uterus rather than a blind steel hook.

I tried to palpate her abdomen, very carefully. It was painful for her. Then I felt a swelling on the right side of the abdomen, near her back, about the size of a large grape. It was excruciating when I touched the swelling.

"She has something going on in the area where her right ovary was located," I said. "I think I should open her up and see what is there."

Marilyn agreed to an exploratory. I was unsure of what could be wrong. This was a pretty small kitten. I hoped I could solve the problem.

In surgery, I found a significant accumulation of fluid along the back. It was retroperitoneal or behind the lining of the abdomen. I aspirated the fluid and explored the area carefully. The fluid had to be urine, but why was it there following a surgery? I investigated again, looking around the kidney carefully and the ureter in the area. I could find nothing more.

We recovered Dorie, and she was like a kitten again with the fluid gone. The area was not painful, but I would need to check her in the morning.

"I need to take her home for the night," Marilyn said. "It has nothing to do with this clinic. It is just that she has been through so much in her young life. She needs some cuddling. I will have her back here the first thing in the morning."

I did not have a restful night. I was going over Dorie's surgery in my mind most of the night. If the fluid was urine, the surgeon who did the spay had to have ligated the ureter, that small tube that runs between the kidney and the bladder. If that was the case, the swelling would be there again in the morning.

I had never heard of a veterinary surgeon making that error. I am not even sure it is listed as a possible complication to that surgery. I did have a niece who suffered that very injury during a hysterectomy some years ago.

In the morning, Dorie was painful again. And the swelling had returned.

"This swelling is back," I said to Marilyn. "I have been up all night thinking about this problem. They had to have ligated her ureter, the tube between the kidney and the bladder. I can send you to Eugene for an ultrasound and some x-rays with a dye that will pass in the urine. We can confirm the problem."

"Listen, Doc, I am a bleeding heart, and I love this little girl, but there are limits to what I can spend," Marilyn said. "Is there anything else we can do?"

"We can remove her right kidney," I said. "If I am correct, that will solve the problem."

"If that is what it takes, let's at least give her a chance to live," Marilyn said.

So little Dorie went back to the surgery table. It was not a complicated surgery to remove the kidney. I also removed the ovarian pedicle where the ovary had been ligated. In doing so, I found the ureter in the mass. That confirmed my suspicion and allowed me to ligate the ureter so there would be no back leakage from the bladder.

Dorie recovered well, and despite having suffered through three surgeries in the last week, she was back to being a kitten. I saved the tissues in formalin, just in case I would need to have a lab confirm my findings.

"I want to make sure that the veterinarian who did this to Dorie never does it to another kitten," Marilyn said in a firm voice. "Can we do that?"

"We can get it investigated," I said. "I am concerned that the humane society might have a charlatan doing their surgeries. I will send them a letter of inquiry and send a copy of that letter to the state examining board."

"Thank you," Marilyn said. "Do I need to do anything special for her?"

"I am sending you home with some antibiotics just to be safe. Other than that, just make sure Dorie is acting like a kitten," I said. "She should be able to live a full life with one kidney. It might become an issue at the end of her life, but that should be a long time."

***

I composed a letter to the humane society and forwarded a copy to the examining board. I had no idea what would happen from that letter. In my letter, I explained Dorie's problem. I expressed a need to determine if this was a surgical error or if there was a need for additional training for the surgeon.

It was a few days, and I got a call from the veterinarian who did the surgery. He was in his mid-seventies and working part-time doing surgery for the humane society. He had no problem saying it was his mistake. He was uncomfortable with doing surgery on these little kittens.

"That will be the last surgery I do on these little kittens," he said. "I don't agree with the practice anyway, so this is going to be a good excuse for me not doing any more spays and neuters on the young kittens."

At about the same time, the investigator from the examining board called to gather my opinion and get information beyond my surgery records. He was most concerned about the age of the surgeon.

After the examining board completed their investigation, they determined there was no need for remedial action against the veterinarian. It was considered a surgical error. The veterinarian was not going to do any more early spay and neuters. The practice of early spays and neuters was not addressed and probably is still an unsettled topic.

***

"Marilyn, I talked with the veterinarian. He is not going to do any more early spays or neuters," I said. "The examining board completed their investigation, and they are leaving it as a surgical error. They are not recommending any remedial action."

"That is a little less than I was hoping for," Marilyn said. "But at least they investigated it. Are you satisfied with their findings?"

"I think so. I think the veterinarian expressed enough remorse to me that he is not going to do any more young cats."

Colleagues

I looked at the large black tumor on Dr. Walker's old gray mare as I wrapped the tail. It was a good thing that the horse was gentle. There were no facilities, and I was at considerable risk, standing directly behind the mare. This tumor was the size of a small egg and located on the right side of her vulva. The lucky thing was it off to the side enough that I could remove it without disrupting the structure of the vulva.

After doing an epidural for anesthesia, I scrubbed the area and soaked it with Betadine. I had the tail tied to the side with a twine around the mare's neck. I palpated the tumor to make sure it was as superficial as I suspected.

Removing the tumor was easy. I made a wide elliptical incision and took a sizable, deep margin. I laid the tumor on the surgery tray and closed the wound with two layers.

"I will send this in just to check on its malignancy," I said. "There is probably not much else we can do, but it will be good for you to know."

"I know," Dr. Walker said. "She is an old mare, but we love her."

"This is a melanoma, for sure," I said. "Black tumor on an old gray horse is almost a description of a melanoma. I was always taught to cut early, cut wide, and cut deep. In this business, the initial surgery is probably our only treatment for these old horses."

"How much do we owe you?" Dr. Walker asked.

"No charge," I said. "I never charge a colleague, something I learned from Dr. Craig."

"That's not fair, I can't make this much up to you," Dr. Walker said.

"My goal in life is to have others owe me," I said. "That way, I know that I have been doing good in my life. I have no expectation of repayment."

The tumor was a melanoma, but not highly malignant. It would not have any influence on the mare's longevity.

***

Some months later, Althea brought in a feral tomcat with a rotten mouth. It was a Friday evening, just after we closed.

The tomcat growled and hissed when I looked into the carrier. I could see raw tissue under his tongue and in the back of his mouth.

"How long have you had this guy, Althea?" I asked. "This is a bad case of stomatitis, and he has some teeth about ready to fall out."

"I have been putting out some canned food for him for several weeks," Althea said. "His mouth is very painful when he eats. It has taken me this long to get him into a carrier."

"Maybe I will try to get an injection into him for tonight and have you bring him back in the morning," I said. "I am not sure I want to keep him in the clinic overnight."

Feline leukemia virus infection was prevalent in Sweet Home. This kind of mouth was one of the presentations we see in cats with FeLV.

"Do you think you can do that without getting bit?" Althea asked.

"We will find out," I said as I worked a snare around the tom's neck.

Once I had him snared, I pulled him out of the carrier and gave him an injection of amoxicillin under his skin.

As I directed his head back into the carrier, he exploded. Up and down and around, he bounced on the end of the snare. When he calmed for a second, I grabbed him by the back of the neck with my left hand to pin him to the exam table.

Now, the fight was on. There was no way he was going back in that carrier, and there was no way I could let go of my grip on him. The ketamine to sedate him was in the lockbox. I called Sandy to get the key from my pocket and get me a dose of ketamine.

When she had a dose in a syringe, I let go of the snare for a second to give the injection. Ketamine burns when it is given in the muscle. Before I could drop the syringe and grab the snare again, this old tomcat turned and bit me on my index finger's knuckle on my left hand. He held onto the bite just to let me know that he had won the battle. I waited, thinking that one bite wound was better than three. Finally, the ketamine began to soak in. I could feel his muscles relax, and the pupils of his eyes dilated widely.

I carefully removed the rotten canine tooth, his right fang, from the wound on my left hand.

"It looks like I am going to hang onto this guy for the next ten days," I said to Althea. "I doubt if he has ever had a rabies vaccination, and if you take him home, he will be gone at the first opportunity."

"Do you think he is going to be okay?" Althea asked as I started to tend to the bite wound he had just inflicted on my left hand.

"My guess is he is a feline leukemia cat," I said. "That probably means that he is not going to do well. But I can check him out in the morning. Right now, I am going to take care of this hand. I will give you a call in the morning."

I should have gone to the doctor that evening. But I flushed the wound several times with saline and betadine. I started myself on some antibiotics off of my shelf.

***

In the morning, I woke with a throbbing hand. My whole hand was swollen, and a lot of pus was draining from the wound. I called Dr. Gulick and arranged to meet him in the ER when he finished his morning rounds at the hospital.

One look at my hand, and I was promptly admitted to the hospital.

I have worked on almost every dangerous critter around, and a damn tomcat puts me in the hospital.

"What am I going to do with that cat?" Sandy asked as they settled me into a bed and started hooking me up to an IV.

"You get Dixie to help you put food and water in his kennel," I said. "Anything more than that can wait."

The culture came back as staph, and they moved me into isolation. My hand was feeling a little better with the IV antibiotics. About then, Sandy called on the telephone. Answering the phone is a real challenge when you have a couple of IV lines on one arm and the other hand in a hot pack.

"The cat died," Sandy said.

"Good," I said. "I don't think I liked him anyway."

"It's not funny," Sandy said. "What am I supposed to do now?"

"You call Dr. Walker," I said. "I am certain she will take care of things. We need to send the head in for rabies testing."

I had no more than hung up the phone, and Dr. Gulick entered the room.

"The cat died," he said.

"Yes, I was just told," I said.

"With you in here, how will we get the cat submitted for rabies testing?" Dr. Gulick asked.

"Sandy is going to get Dr. Walker to take care of things," I said. "I don't think there will be any problems."

***

By Monday morning, my hand was doing well, and I was released to get back to work.

I called Dr. Walker to thank her for taking care of things.

"I just wanted to call and say thanks for helping Sandy with that old cat," I said.

"Are you doing okay?" she asked. "Those cat bites can be bad, and the mouth on that guy was as rotten as I have seen."

"I was a little worried on Saturday morning," I said. "But the IV antibiotics and the hot packs did the job, I guess. We will probably hear about the rabies results today. I think that cat was probably a feline leukemia cat."

"Well, I hope it is not rabies positive, for both our sakes," Dr. Walker said.

"So, how much do I owe you for your work?" I asked.

"How much do you owe me!" Dr. Walker said. "Are you kidding? You come all the way up to my place and stand behind our horse to do surgery, and you say, "No charge." You have to be kidding. You don't owe me anything. We are colleagues, remember."

Don't Put Her in the Barn

Looking around as I waited for George, it was apparent this is a well-kept old ranch, probably by a perfectionist. There was nothing out of place. The barn was old, with a bit of bow in the roof's ridgeline, but it had a fresh coat of bright red paint. The small white ranch house sat in the middle of an immaculate yard, with a white picket fence and rose bushes galore.

The old cow in the corral looked like she had lived many years here also. She was a cow that I would have recommended culling from the herd years ago, had I been asked. Her udder anatomy had to have been an issue for many years. Her two hind teats were large and hung low, almost dragging the ground, making them almost impossible for a calf to nurse. By the time the calf nursed those teats, the quarters would be devoid of milk.

The incidence of mastitis in cows with this type of anatomy is high; virtually all of them will have problems if they live long enough. I would guess that is the problem I am going to look at today. Even though George just said a sick cow when he called.

I got out of the truck and walked over to the corral fence to look at the cow a little better. I could see George putting on his shoes on the porch of the ranch house. Here he came on a trot.

"Hi, Doc, I'm sorry I made you wait," George said as he extended his hand. "The Mrs. had me hanging pictures of the great-grandkids."

"That's not a problem, George," I said as I shook his hand. "This is the end of the day, and I have plenty of time. I am betting that this is the sick cow. And I am betting that she has mastitis."

"You're correct on both bets," George said. "You can't see from here, but she has a black teat on the back far side. And she is pretty sick; she stands there and doesn't want to eat or drink. She hasn't worried about her calf at all."

"Did you just notice her today?" I asked.

"She was fine yesterday," George said. "But you know, she has been a thorn in my side every year for the last several years. I have a devil of a time getting her calf hooked up on those back teats. I know you will probably tell me I should have sent her down the road a long time ago. But you know, she always weans one of the best calves in the bunch. Some of these old girls earn their hole in the ground."

"Now you're right on both counts, George," I said. "I would have told you to cull her years ago. And because she pays you for your extra efforts with a super calf every year, she probably does deserve her hole in the ground."

"Do you think you can do anything for her?" George asked.

"Let me get a few things and get a look at her," I said. "Do you think I need a rope?"

"She hasn't moved a muscle in the last hour," George said.

Her problem was easy to see when I walked around the cow. Her right hind teat was black and cold to the touch. The discoloration extended up the backside of the quarter. Here was a case of mastitis with a dead quarter. Probably an acute E. coli mastitis, the circulation is disrupted by the infection, and the tissue dies. The cow will die unless we can get the disease under control.

"I am going to have to cut this teat off, George," I said. "And maybe open up this quarter more than just the teat."

"Isn't that going to bleed a lot?" George asked.

"No, this tissue is all dead," I said. "The only chance we have of saving this cow is to get some drainage out of this quarter, put her on some antibiotics, and hope for the best."

I took a scalpel and cut the teat off. A lot of fluid drained out of the quarter, and some tissue hung out of the hole. I gave a little tug to the yellowish chunk of tissue hanging out of the gap left by the missing teat. A large mass of dead mammary tissue plopped out of the hole. With my gloved finger, I was able to pull another two chunks of tissue out of the quarter. I flushed the quarter with hydrogen peroxide and followed with Betadine.

"I will give her some long-acting antibiotics so you won't have to mess with her in the morning," I said.

After treating her, I put things away in the truck and explained to George how the tissue in that quarter was dead and that more chunks would fall out of the large hole where I removed the teat.

"If you see stuff hanging out of that hole, you need to pull it on out of there," I explained. "Otherwise, it will just block the hole, and we will lose the drainage."

"Do you think she is going to be okay?" George asked.

"We are just going to have to see what morning gives us," I said as I got into my truck to leave.

"Do you think I should put her in the barn tonight?" George asked.

I looked at the barn. The door on this side of the barn was open, and all I could see the inside of the barn was a maze of small pens. It must have been an old sheep barn.

"If she dies, can you get her out of there easily?" I asked.

George looked at the barn and thought for a moment. "No, I don't think I could get her out of there very easy at all," he said.

"Don't put her in the barn tonight," I said as I pulled away.

***

The old cow did live through the night. The entire quarter fell off eventually, and it did heal up. It didn't look good for a time. The old cow raised another calf the following year before finally finding that hole in the ground.

All Hell Broke Loose

"I have been looking at Sophie's horse for a couple of days now," Don said as we were lining out our morning calls. "I would like you to come along this morning and get your opinion. This is low-grade colic. I think it must be an obstruction, but I have not been able to confirm it."

The three of us in the practice in Enumclaw were cow doctors. We did fine with the small animals, but horses were a different story. We had a few horse doctors around, and we tried to send most of the horses their way, but a few of the dairies would have a horse, and we were usually stuck with taking care of them.

"I didn't know they had a horse," I said.

"She bought it a few months ago," Don said. "Harold says he isn't going to spend any money on it, but you know how that goes."

We pulled into the barnyard in Don's truck. The horse was in an open shed across from the milking parlor. She was standing, head down, and kicking at her belly every minute or two.

"Have you done a rectal exam?" I asked Don.

"Yes, but I didn't find anything," Don said. "But I am not sure about the horse's gut. That's why I wanted you to get a look. Your schooling is a lot more recent than mine."

I started a complete exam just so I wouldn't miss anything. All her vitals were normal, except her pulse was a little elevated. Then listening to her abdomen, there were virtually no gut sounds.

"I guess I should do a rectal exam," I said. "I hate doing one on an unrestrained horse, but I don't think we are going find a stock here."

"She was pretty good for me," Don said. "I will hold her for you."

"Maybe we should put a twitch on her," I said.

"Don't let Sophie know about that," Don said. "She was pretty protective of her when I did the first exam."

"If she says anything, just explain that a rectal exam on an unrestrained horse is dangerous to the vet but also dangerous to the horse," I said. "If she jumps or kicks, she could end up with a ruptured colon, and that's probably fatal."

I wrapped the tail to keep the hair out of the way and lubed my OB sleeve with a large amount of J-Lube. Standing on her right side, I held the tail to the side with my right hand and slid my left hand and arm into her rectum. She stood still.

The colon was utterly empty. Don was probably correct in this being an obstruction colic. Pushing my arm in deeper, I explored the gut that was within reach. I found nothing in this initial exploration.

I reached deep and down on the left side of her abdomen, searching for the pelvic flexure, that spot where the large colon of the horse does a u-turn. It is the smallest portion of the colon and is often the site of obstructions.

Finally, the pelvic flexure almost jumped into my hand. There was a firm impaction in the flexure. I massaged this mass, and the mare danced from side to side. This impaction was not large. I was surprised that the earlier treatments with mineral oil had not moved it. I slowly removed my arm and rinsed her rectum.

"You are right about this being an obstruction," I said to Don as I unwrapped the tail. "There is firm impaction at the pelvic flexure, and it was painful for her for me to try to massage it."

"I figured you had found it when she started acting up," Don said.

"I know the experts say not to mix mineral oil and DSS," I said. "But if you have treated her with oil a couple of times over the last three days, I think we should give her a big dose of DSS. If that doesn't do it, and Harold doesn't want to send her over the mountain to Washington State, I think we should talk to them about doing surgery."

"That's a big step," Don said. "Are you sure about that? I have always been told never to touch the gut of the horse."

"When I was home a couple of years ago, I went on a call with Dr. Haug to look at a horse he had done a flank surgery on for an impaction at the pelvic flexure," I said. "That surgery went well. He just made a left flank approach, pulled out the pelvic flexure, and massaged the impaction to break it down. Worked like a charm."

"I guess we could give it a try," Don said. "I know Harold will go for it. He is sure this horse is going to cost more than they paid for it in the first place."

"When I went back to school after visiting with Dr. Haug, I asked Dr. Adams about that procedure," I said. "He said, if you have no other option, get in and get out as quickly as possible, and you will be fine."

We gave the mare a full dose of DSS and some Dipyrone for pain. As we were finishing up, Harold came out of the house to talk with us.

"What do you two think?" he asked.

"We definitely have an impaction at the pelvic flexure," I said. "We gave her some more medication. If she hasn't passed anything by morning, we think we should do some surgery. Either that or send her over to Washington State."

"We are not going to send her anywhere," Harold said. "I'm not sure about you guys doing surgery either. But you know, Sophie will say we have to do what you say."

"Okay," Don said. "No food, just like before. We will be here, ready to do surgery the first thing the morning."

"I guess we should hit the books tonight," Don said as we were driving back to the clinic. "I am not sure I can find the pelvic flexure. In fact, I have never done a flank incision of a horse."

"Believe it or not, we did several while I was in school," I said. "The incision is a piece of cake. In fact, the whole surgery is a piece of cake. It is just getting it done and not ending up with peritonitis following surgery."

"We will plan for you to do the surgery. I will just be there to learn," Don said.

The following day we spent a lot of time packing and repacking our supplies for the surgery.

"I don't want to be out there and find out we forgot something," Don said.

Finally, all packed up, we made the drive out to Harold's place. When we pulled into the driveway, we could see the shed where the horse was being kept. As we got closer, we could see several boards were missing on the side of the shed. They were lying out on the pasture.

"That doesn't look good," Don said. "I hope the horse is still alive."

When we pulled up to the shed, the mare was standing head up and eating some alfalfa. The inside of the shed was plastered with horse manure. And there were 3 boards kicked out on the side of the shed.

Harold came out of the barn when we arrived.

"About three this morning, it sounded like all hell was breaking loose," Harold said.

Sophie came up behind Harold. "We got up and came out here, and she was acting like everything was fine," Sophie said. "She had kicked out the side of the barn, and there was horse shit everywhere. We figured that everything was okay, so I gave her just a little hay. She loved it."

"This is the best thing that has happened to me in a long time," Don said. "I have worried about this surgery all night."

"What do we do now?" Harold asked.

"Let's just go easy on the feed for a few days," I said. "Maybe plan to be back on a full diet this time next week. Until then, just small amounts of hay and grain and plenty of water."

We got back in the truck, and Don let out an audible sigh. "I am so relieved that we didn't have to do any surgery this morning."

"Makes the rest of the day easy," I said. "Maybe we should stop off for a beer on the way home tonight."

"That sounds good," Don said. "I'll buy."

Driving Blind

This was one of those extra busy days in the middle of July. The appointment book was full. Most of the problems were flea problems. I always became weary of discussing the flea life cycle with folks. We were trying to put out the fire that should have been controlled last winter.

I noticed Ed sitting out in the reception area. He had a small dog on his lap. I knew Ed from my days bowling for the Elks Club team in the Thursday night men's league. Ed was older, probably in his late 60s, balding, and he wore thick glasses. He was short and stocky and somewhat rounded by the years.

"What's up with Ed?" I asked Sandy as she was passing on a run.

"He has a flea problem with his dog, but no appointment," Sandy said. "I told him we would try to work him in, but we were booked solid today."

I stepped back into the exam room and started my rehearsed spiel on flea control. Flea control was difficult to impossible if folks had not been working at it all year round. When the heat of July and

August hit, the fleas hatch from everywhere. They just about eat the poor dogs and cats alive.

"This is a typical case of flea allergy dermatitis," I started. "The hair loss and the skin lesions are back here by the tail. It doesn't take much looking to see the fleas scatter."

I spread the hair on the dog's back, and fleas ran in all directions. I turned him over and showed a dozen fleas on his belly.

"The flea collar doesn't do a lot of good at this point," I said. "His little cloud of protection is a few feet behind him as he runs around the house."

"There are no fleas in my house, Doctor," Mrs. Jones said flatly.

"This time of the year, fleas are everywhere. Getting them under control is difficult. It requires flea bombs in the house and flea dips on the dog. And then, the process needs to be repeated every week for several treatments. Sometimes we need to use a yard spray also."

"I don't have fleas in my house," Mrs. Jones repeated. "We had to sit out in your reception area for almost fifteen minutes today. That must be where all these fleas came from."

"I can give you some medication that will help the skin, but without flea control, Rex is going to have a bare tailhead until winter. If you would like, I can give you a referral to a veterinary dermatologist in Newberg."

At about this time, I hear a big commotion in the reception area. I excuse myself for a moment. I get a brief respite from Mrs. Jones.

Ed is up on his feet and arguing with Sandy and Ruth.

I step out to the reception room. Ed is red-faced and clutching his little dog tight enough that his eyes are bulging a little.

"Doc, damn it, I have been sitting out here for fifteen minutes, and this lady comes in, and they take her back before me. It is my turn."

"Ed, Sandy told you we would try to work you in. We have a solid book of appointments this afternoon. That lady had an appointment. My policy is to see appointments on time if possible."

"Damn it, Doc, I am telling you, it ain't fair," Ed said.

Now I was a little upset. I took a step toward Ed and pointed to the door.

"Ed, the door is right there, don't let it hit you in the ass as you leave," I said. "The next clinic is about nineteen miles down the road."

"No, I don't need to do that," Ed said. "I just wanted to make sure I was not being slighted by the girls. I will wait for my turn."

Ed returned to his seat, and I returned to Mrs. Jones, probably not in the mood to discuss where Rex's fleas came from. I am sure everyone in the clinic had heard the altercation.

"I don't need a referral," Mrs. Jones said. "You can fix me up with what I need, and we'll see how it works."

That was a lot easier than I expected, and it allowed us enough time to get Ed into an exam room.

When I finally got to Ed, we were both apologetic. We treated his little dog for fleas, along with most of the other dogs that day.

***

Several months later, Ed and his wife were in the clinic at a much quieter time. I had ample time for some casual conversation.

"Are you still bowling, Doc," Ed asked?

"No, I have an old football injury to my left knee and figured I better give it up," I said.

"I had to give it up, too," Ed said. "My vision got so bad that I couldn't see the pins at the end of the alley."

"That would make it pretty difficult," I said.

"Yes, and I don't drive," his wife said. "So when he is driving now, we drive the old pickup with bench seats. I sit really close to him, sort of like when we were teenagers, and I tell him what is coming down the road."

"That sounds a little dangerous," I said. "You guys maybe should get some help before you are in a wreck."

"It works okay right now, Doc," Ed said. "Our son is getting his job changed around, and he is going to move his family in with us pretty soon. Things will be better then, and I won't be driving so much."

"Okay, that sounds better, but you need to be careful."

Elk Delivery

"Doc, I have a cow elk that has been walking around the pasture in labor for the last couple of hours," Frank said into the phone. "I can see a small sac of fluid and a foot once in a while, but she's not making any progress."

"If she has been at it a couple of hours, we probably should get a look at her. Don't dart her until I get up there."

Frank McCubbins had quite a variety of exotic animals. Sika deer, fallow deer, some antelope, and a small herd of elk. The elk herd consisted of a bull and five or six cows. There were no facilities for handling any of these. We were stuck with using a capture dart.

"What drug do you want me to use," Frank asked as we were loading the dart.

"She is not too high strung, and with this difficult birth, a dose of Rompun will probably do the trick. We want her to recover pretty quickly so she will take care of the calf."

"She is not too big," Frank observed. "What kind of a dose should we use?"

I had provided a dosage chart to Frank so he would not have to do any calculations on dosage. The chart was set up to give the volume dose in milliliters for each weight in fifty-pound steps.

"Let's give her a five-hundred-pound dose," I said. "She might be a little over that, but not by much. And we want her to recover quickly."

Rompun was a tranquilizer approved for the horse and small animals. We routinely used it at very low doses on cows. It was helpful for short-term procedures that required chemical restraint. Its shortcoming was animals who were flat out could suddenly recover and react defensively.

We stepped through the gate into the elk pasture. The bull and the other cows moved to the far corner. We could usually lure the herd to the feed rack with a bucket of apples, but this problem cow was by herself away from the others. I could see her getting up, turning around, straining, and lying down again.

She did not seem to be bothered by our approach. When we were within twenty yards, she stood up, and Frank fired the dart gun, striking her in the hip with the loaded dart. We moved away to allow the drug to take effect.

Once she was on the ground and her head turned to her side, we approached cautiously.

"Let's get a rope on her just in case she jumps up when I start working on her," I said. "We don't want to have to give a second dose."

Jim, Frank's hired man, placed the lasso over her head and backed away, holding the rope with gloved hands just in case she came alive.

I removed the dart from her left hind leg and applied some Betadine to the wound. Then I washed her rear end and carefully explored her birth canal. She had no response.

I could feel one front foot and then the nose of the calf. I reached deeper. The second foot was not in the birth canal.

"The calf has a front leg back," I said. "I should be able to get it up into position easily. There is plenty of room in there. He will pop right out once the position is corrected."

I reached in and ran my hand past the shoulder on the calf and along its side. I grabbed the cannon bone on the retained leg and pulled it forward. Then I flipped the hoof up into the birth canal.

The cow elk raised her head when I corrected the leg position of the calf. I grabbed the front feet and pulled. The calf quickly slipped out onto the ground. I pulled the calf around toward the cow's head and stood up.

"Go ahead and remove that rope, Jim," I said.

As soon as the rope was off, I gave the cow a slap on the butt. She instantly jumped to her feet.

At the same time, she kicked with a hind leg. The kick was directed with deadly accuracy at me. She walked away, going only a few steps up the hill.

I was lucky that I was just far enough from her that her hoof only brushed my shirt. I had a neat imprint of the toes of her hoof on my shirt just above my navel.

"I think that would have hurt a little if I had been an inch or two closer," I said.

"I would guess so," Frank said, with an expression of concern on his face.

"Let's just move away, quietly, down the hill. Hopefully, she will return and take care of this calf," I said.

I quickly gathered my stuff into my bucket, and we moved down the hill. Looking back, she was watching both us and the calf. By the time we got to the gate, she was back licking the calf.

"Frank, I think we got a little lucky," I said as we opened the gate and left the pasture. "You want to write that dose down on your chart. That worked perfectly."

"I'll try to remember to do that when I get back to the house," Frank said. "Is there anything I need to do with her now?"

"I don't think so," I said. "She would have probably been coyote food in the wild. There would have been no way for that calf to come out on its own."

I watched up the hill as I turned my truck to the gate at the end of the lane. Momma elk was tending to the calf. Things were going to be okay.

Fleeing the Flea

It is another hot August day in Sweet Home, and fleas are eating most of the dogs alive.

Returning from lunch, I could see that Dixie had started the sprinkler on the roof already. That helped keep the clinic cool. The water would run hot, coming off the roof.

Joseph was waiting with a worried look on his face. He was holding a limp Domino. Domino was a little five-pound Chihuahua. He was black as a young dog but was half gray now.

Dixie herded Joe and Domino into the exam room as soon as I put on my smock.

"He is not doing so good this morning, Doc," Joe said. Joe was in his early seventies and had lost his wife several years ago. Domino was about all he had left in the world.

We placed Domino on the exam table on a fleece. I pulled up his lip. His membranes were white. I looked at his lower back and ventral abdomen, and he was literally covered with fleas.

"Joe, Domino is being eaten alive with fleas," I said. "I am going to run a quick CBC on him. I can see that he is anemic. We just need to know how bad."

After we drew a small tube of blood, I discussed the flea situation with Joe as I waited for the results. The in-house blood machine would only take a few minutes.

"I don't know why he would have so many fleas," Joe said. "He has had a flea collar on since the first of the month."

"It is a little complex, Joe," I said. "You have probably had fleas laying eggs in the house all winter and spring. That flea collar might work a little around his head and neck, but for the most part, that little cloud of protection is about three feet behind him. When the weather gets hot, all the fleas come alive, and for a little guy like this, they suck the blood right out of him."

The CBC showed a packed cell volume of less than six percent. I don't think I have seen a level this low in a living dog.

"Joe, Domino is very critical," I said. "I need to get some blood into him right now. Any undue stress and he could drop dead in an instant. We will need him for a couple of hours, and I will talk about what we need to do when you pick him up."

Luck was on Domino's side. Riley was in the clinic today. Riley was a large mixed-breed dog weighing over one hundred pounds and would make a suitable blood donor. I got ready to collect blood while Sandy called Riley's owner.

"We have an emergency with a little Chihuahua. We need to give him a blood transfusion," Sandy said into the phone. "We only need an ounce or two of blood and would like to collect that from Riley, if that is okay. That is a small enough volume that Riley won't miss it."

They consented, of course, and I drew the blood into a heparinized syringe. Then we turned around and administered it to Domino via a jugular catheter. The risk of a transfusion reaction on an initial transfusion was low, and Domino's blood values dictated immediate blood.

The result was almost instantaneous. Domino came alive again. His membranes pinked up, and he sat up and looked around as if to ask, "Where am I?"

I gave Domino a Capstar tablet. This was a new pill that provided close to a total flea kill in thirty to sixty minutes. I also gave him some oral prednisone to reduce the inflammation in the skin.

When Joe returned, we had him fixed up with some topical Advantage for flea control, and I spent some time discussing year-round flea control. We would have needed to use a flea bomb in the house in the old days, but those were almost impossible to find now. The newer products did a good enough job that we did not have to treat the home.

"The important thing to remember is to maintain flea control all the time, year-round," I said. "In August, when it turns hot, I probably spend ninety percent of my time treating dogs and cats with skin issues. And most of those issues are caused by fleas."

"Okay, Doc, I don't want to lose this guy," Joe said. "I would have never thought that fleas could do that to a dog."

"It all depends on the dog," I said. "Domino is not much of a dog compared to Riley, his donor. Riley weighs over one hundred pounds, and fleas could not do that to him. But Domino should be okay now, you just bring him by next week, and I will recheck that blood, just to make sure he is doing okay."

Joe left with Domino in the crook of his elbow, Domino standing on his front feet, trying to lick Joe's face, one happy ending.

Dixie had the next patient ready in the exam room, an older lady I had not seen before. Doris had a poodle, Daisy, who was scratching on her tail head, that area on the low back above the tail. This was the textbook appearance for flea allergy dermatitis.

"Daisy has been scratching herself raw," Doris said.

I looked Daisy over from head to tail. Everything looked fine except for the skin. She had just come from the groomer. I ran my hand over the sparse hair on her low back—fleas scattered in all directions.

"Doris, this pattern of hair loss is what we see with flea allergy dermatitis," I said. "We need to use some medication along with some flea control, and this will clear right up."

"I overheard your conversation with the gentleman who just left," Doris said. "I want you to know, Doctor, Daisy does not have fleas, and there are no fleas in my house. The groomer thinks this is a food allergy."

I promptly parted the hair on Daisy's back again and quickly captured a flea. I placed the flea on the exam table and squished it with my thumbnail. I didn't say a word.

Doris did not change her position. This was a food allergy.

"We just happen to have a new veterinary dermatologist that has started practice in Eugene," I said. "She would be the one who you should see to handle Daisy's possible food allergies. I will send your records down to her and send you home with her telephone number. I think that she will be able to get you right in to look at Daisy since she has just started practice."

Doris and Daisy left with the information.

"That was quick," Dixie said.

"I am too tired to spend my time talking to a brick wall," I said. "The dermatologist can tell her it is a flea problem after she does five hundred dollars worth of skin testing. I am sure she will believe her then."

Harry and Buffy

Harry's car pulled up to the front of the clinic and bumped hard into the curb in the diagonal parking space, one wheel of the vehicle almost coming up on the curb.

Watching out the window, Joleen said, "It looks like Harry has been drinking too much again."

Harry stepped out of the car, and you could appreciate how large a man he was. He just seemed to keep coming. He steadied himself a bit, with both hands on the roof of the car, and leaned in to pull Buffy off the passenger seat.

Harry was an older man, well into his seventies, if not eighty. He lived by himself. The story goes that he started drinking heavily after his wife's death, almost ten years ago.

"I wonder what Buffy has been up to this time?" Joleen said.

Buffy was one of those dogs who could be termed a mutt on his medical record, and he would fit the bill. If one had to pick a breed, you would probably call him a terrier. Small and rugged, he was not much to look at, but he was intensely loyal to Harry. He was perhaps the one thing that kept Harry from going off the cliff with his drinking.

Buffy was also a tough little guy and would take on the biggest dog on the block every chance he could get. We had sewn up more than one gash on his body. His thick bristly hair coat hid most of the scars well.

Most of the time, when we would see Buffy, Harry was drunk. Sometimes, almost falling down drunk. It often took Harry several days to remember where he had left Buffy, as is often the case when the owner needs someone to watch after him.

Buffy was always protective of Harry and his space. Most of his visits came from wounds received in dog fights: bite wounds, broken legs, and various scrapes and bruises. Harry somehow always paid the bill.

Harry came through the door holding Buffy with bloody hands. He immediately handed Buffy to Joleen.

Joleen looked at the wounds and gasped. "What in the world happened to Buffy this time."

"Two big dogs got him. They bit me, breaking up the fight. They were going to kill him this time."

Buffy had deep punctures on both sides of his lower back and extensive muscle and skin damage.

"Harry, we will take care of Buffy," I said. "You need to go get a doctor to look at that hand. Do you have somebody we can call to drive you there?"

"Yes, I have already called Jim to come to pick me up," Harry said. "I think he just pulled up."

"What should I tell Harry about how long will we be keeping Buffy?" Joleen asked as she started helping Harry out to the waiting car.

"Don't worry about it. We will know more about how Buffy is going to do by the time Harry remembers where he left him," I reply.

Buffy's wounds were a real challenge, and had he not been so tough, he would not have survived. By the third day after admission, we could recognize extensive tissue death in the area of his wounds.

We went through a series of three or four surgeries to remove dead skin and muscle. By the time we had all the dead tissue removed, Buffy had lost a significant portion of skin and muscle on his left side and hip.

Buffy spends twenty-one days in the clinic, and he hated every day of it. One could hardly blame him. Two or three injections and the constant bandage changes must make him believe we exist only to torture him. He cowers every time he sees me.

He is ecstatic when Harry finally takes him home. He still has large open wounds, but they are healing well, and finally, I believe, the wounds can be managed by Harry at home.

On the fourth day after Buffy was home, Harry calls the clinic. He's drunk, but he can still talk.

"Buffy's sick, can hardly walk." Harry finally stutters into the phone.

Not sure who could hardly walk, Joleen asked, "Can you get him to the clinic, Harry?"

"Don't think I can drive much right now," Harry replies, with a stroke of insight that is uncommon for him.

"We will pick him up right after lunch, Harry. I just need to know where you live."

I have received many different sets of directions in my years of practice. I have often criticized women for what I perceived as a failure to pay attention to details and inability to get accurate directions that a person could follow. But Harry's directions were impossible.

Despite those directions, Joleen and I pulled into his driveway shortly after lunch. Harry lived in a small run-down shack, but it was surprisingly well kept.

We knocked on the door, and in a few minutes, Harry opened the door. He was hooked up to his oxygen bottle and having a little trouble walking. Buffy was at his heels. When he looked up and saw us, he had absolute dread in his eyes.

"My God, they know where I live," those eyes seemed to say. Buffy reared back and headed for the back room, staggering on stiff legs. He was attempting to crawl behind the small cabinet when I caught up with him.

"What's wrong with him?" Harry asked.

"It looks like Buffy has tetanus," I said. "Tetanus in the dog is rare. I have only seen it in one other dog. The good thing is dogs are resistant to the disease, and most will survive with treatment."

Joleen took Buffy from my arms, "I think he feels safer with me."

"We will probably need to keep him for another week or two, Harry. We will give you a call when he is ready to go home," I said.

Buffy spent another twenty days in the clinic. He responded well to treatment. We kept him a few extra days to make sure Harry could handle his treatments at home.

This time, we had Harry bring Buffy to the clinic several times a week. Just so we could keep track of the wounds. These visits became a struggle. Buffy would be under the car seat before Harry was fully parked in front of the clinic. Joleen had to wrestle him out from under the car seat and into the clinic.

"Harry, next visit, you call when you leave the house, and you park over at Safeway," Joleen instructed. "I will come over there and get Buffy."

On the first trip after that, Joleen opened the passenger door and grabbed Buffy before he could get off the seat. Harry staggered his parking location on each visit, and Buffy never seemed to catch on to the game.

Finally, Buffy's wounds healed. He was scarred but functional.

"Now you just have to keep him from going out and picking a fight with the big boys," I told Harry as he made his last visit.

"I think this little guy is going to be an inside dog from now on," Harry said. "I will probably have to stop drinking. That is what got him into trouble last time. I let him out to do his business because I was too drunk to walk him."

"Maybe both of you have learned a lesson," I said. "It will be a good thing if Buffy helps you to slow down on the bottle."

"How much do I owe you, Doc?" Harry asked.

"Your bill is pretty big," I said.

"I don't have much, but I will pay you fifty dollars a month, probably forever," Harry said as he shook my hand.

Harry faithfully paid fifty dollars a month, every month until he died. He was always thankful for Buffy's recovery. If people were half as sincere as Harry, credit problems would be non-existent.

Buffy hated me for the rest of his days.

The Berserk Mule

Life is supposed to get better when the kids are grown and have families of their own. When you more or less retire and enjoy the fruits of your labors. That was what Billie envisioned.

Then her son and his buddy bought a mule, and they needed a place to keep it. That was an easy decision to make since Mom and Dad had a small ranch.

"We can keep it out at the folks," the son says to his buddy.

It was not an impressive-looking mule. It may have been a hinny, but they believed it was a mule. It was small in stature compared to most mules that I have known. But it had a good history as a pack animal.

They purchased it just for that purpose, to use as a pack animal for their hunting trips. So, it had nothing to do for eleven months of the year except to establish dominance over the cows, which reluctantly shared his pasture.

Everything was fine for the fall and winter. The little mule moved about the farm with no problems. He seemed to get along with the cows fine. He could be observed biting at a cow once in a while to make sure she did what he thought was correct. But other than that occasional bite on a cow to make sure she was doing what the mule expected, there were no issues.

Spring came, and the cows were calving. As the numbers of calves increased, so did the stress on the mule.

"I need Doc to come now!" Billie said into the phone as soon as Sandy answered. And then the line was dead.

Sandy luckily recognized Billie's voice and tried to return the call but got a busy signal. We had no idea what the emergency was, but it was obviously an emergency. I jumped into the truck and headed toward Pleasant Valley.

When I arrived, Billie was in the driveway to the barn, turning circles with a small shotgun in her hands.

She grabbed me by the arm and rested her forehead on my shoulder, with a big sob, she says, "I am so glad to see you, Doc."

"What's going on, Billie?" I asked.

"That damn little worthless mule has gone berserk," Billie said. "He has been attacking the calves. Not little nips like he did with the cows. He was picking these calves up by the back of their neck and throwing them in the air. Then he was falling on them with his knees. I know he was trying to kill them."

I glanced out in the field by the barn. All the calves were up and moving with their mothers. The cows milling around in wide circles, obviously upset.

"I am here all alone," Billie said. "I had no idea what to do. I called your office and then decided I needed to shoot the worthless thing. We have a whole house full of guns, and I don't know a thing about any of them."

I reached out and took the small shotgun out of her hands. If she didn't know anything about guns, it would be safer in my hands while she was telling the story.

"I grabbed this gun, and then I didn't know what shells fit it," Billie said.

I looked at the gun. It was a single shot four-ten. I opened the breach, and there was a thirty-thirty shell in the chamber. It had been fired.

"Is this what you used to shoot at him?" I asked.

"Yes, that is the only thing that I could find the would fit into the thing," Billie said. "I came out here and pointed it at him and pulled the trigger. I don't think I hit anything, but he must know a gunshot. He stopped right now. I was trying to decide what to do next when you pulled into the driveway."

Billie grabbed my arm again, "I was never so happy to see somebody."

"Billie, this is a rifle cartridge that you shot in a shotgun," I said. "You are probably lucky this little gun didn't blow up with you."

"It was the only thing that I could find that fit," Bille said.

"I will go out and check the calves," I said. "I guess I should check the mule also, just to make sure there isn't a bullet hole in him somewhere."

"You can look at him, but we are not going to spend any money on him," Billie said. "The boys are just going have to find another home for him. He is done at this place."

"I will walk through the calves and just make sure there are no major injuries," I said. "They all look good from here. I will run the mule into the barn and close the gate. You don't need to have any more excitement this afternoon. When will Bill be home?"

"Bill is out of town for a couple of days," Billie said. "I have Larry called, and he will come out when he is off this afternoon. But if you can get that jackass into the barn, I would appreciate it."

I walked through the calves. The cows were all upset and reluctant to give me much access. Everyone was moving well and looked okay. There were a couple of scrapes on the back of the neck on a couple of calves. There was no way to deal with that this afternoon.

The mule was a sucker for a can of grain, and he followed me into the loafing shed side of the barn. I was able to get the gate closed and latched. He was not happy when he realized that he was trapped in there. He just didn't understand how lucky he was that Bill wasn't home. Bill was a crack shot, a marine veteran who fought in the Pacific during WWII. He would not have missed.

"I think the calves are okay," I said as I returned from the barn. "There are a couple of scrapes, but I think those will heal without any treatment. The mule is secured. He might need some water. You might have Larry check that when he gets here."

"It is the other boys who I am going to call right now," Billie said. "They are going have to be out here tonight or tomorrow and move that guy somewhere. His welcome here has expired."

Hot Tub Skin Infection

"Doc, this is Dave. I just brought in a cow from the back pasture," Dave said into the phone. "I think she has a dead calf in her. I have been busy, and it has been several days since I checked those cows. I guess I didn't even suspect this gal was pregnant. But she is sloughing a lot of fluid and smells pretty bad."

"I can get up there this afternoon if that works for you," I said. "Do you have the cow in the barn?"

Dave had the cow in the crowding alley when I pulled up to the barn. I enjoyed going to Dave's place. He had a commanding view of Sweet Home, plus his facilities for working his cows were some of the best around.

This was a big Red Angus cow. I could smell her when I stepped out of the truck.

"The way she smells, that calf must have been dead for several days," I said.

"It's been several days since I checked that pasture," Dave said as he loaded her into the squeeze chute.

This was one big cow. Reaching the depths of this girl's uterus will be impossible for me. With the volume of fluid she is discharging and the odor, this would be a real mess. After scrubbing her well, I pulled on a plastic OB sleeve on each arm.

I ran my left arm into the vagina. I ran into a shoulder of the calf lodged into the birth canal. Feeling around, the head is turned back to the right side of the calf. Both front legs are retained. I could feel the hair slipping off the calf while I maneuvered my hand around him while exploring his position.

When I pulled my arm out, the sleeve was covered with black hair.

"Dave, this calf has been dead five to seven days to be losing hair like this," I said. "This is going to be a mess. I think the best thing for the cow is to do a fetotomy. I always try to do a fetotomy on a dead calf first. I think we see better fertility in the cow following a fetotomy versus a C-section.

"That sounds good to me," Dave said. "What do you need from me?"

"Aw, actually, you are going to have to do a lot of the real work," I said. "You are going to have to do all the sawing. I will have to position the wire saw and hold the fetatome in position while you do the work."

With a fetatome, I could make right angle cuts on the fetus. My first cut would be to remove the head and neck. That should allow me room to bring the front legs into the birth canal. Then it will depend on how much air has accumulated in the abdomen of the calf.

Passing a wire around the neck of the calf proves nearly impossible. I finally have to strip down to my waist and go in with a bare arm to drop an OB chain over the top of the neck.

"Dave, I need something to stand on," I said. "This big cow is a long reach for me."

Dave brings a big block of wood, an oak round, for me to stand on. Standing higher, I reach as deep as I can. My shoulder is in the vulva now. Finally, I grab the chain on the underside of the neck.

Tying the OB wire saw to the chain, I can pull the wire around the calf's neck. Then I thread the wire through the two barrels of the fetatome. This fetatome sort of looks like a cross between a double-barrel shotgun and a trombone.

With everything in position, I give Dave a brief lesson on how to run the saw handles. He is a strong young man. This won't take long.

"The only thing I want you to remember is my hand is holding the end of the fetatome in position," I said. "If you hear me holler, you stop. That saw will take a finger off in a single pull."

Severing the neck takes less than a minute. I quickly set the fetatome aside and run my arm back in to grab the head, no luck. The head has slipped into the depths of the uterus.

I can reach the front legs, but I had to stand on the woodblock to accomplish that feat. I have little difficulty in pulling the legs into the birth canal.

I mix some powdered J-Lube with half a bucket of water and pump this into the uterus to provide some lubrication. With the head gone and the lube, this calf pulls out with little difficulty.

Now the only thing is to get the head. Reaching as far as I can into the uterus, I can only just touch the head. I try again and again. There must be another way.

"In school, they always stressed time," I say to Dave, as much for my benefit as his. "If you haven't accomplished what you are trying to do in twenty minutes, you better be doing something else."

"So, what else is there at this point?" Dave asked. "Are you going to do a C-section for the head?"

"That would be like doing a C-section for the last puppy after spending half the night to deliver the first 10 pups," I said. "I have been there, done that. If I can't get it out, we might have to do that. But first, we are going to use some tincture of time. I will put a package of tetracycline powder in this uterus and load her up on some antibiotics. I will recheck her in the morning. Hopefully, this uterus will shrink up enough that I can get hold of the head and pull it out."

"You think she will be okay?" Dave asked.

"I think so," I said. "The cow is a funny beast, though. This calf has been dead in there for a week, and she is looking pretty good. Then I come and dig around in there, and it knocks a lot of bugs off into her bloodstream. We have to load her up on antibiotics. Otherwise, she will be in a problem in the morning. The biggest risk with this plan is if the cervix closes up too much. Then we might not be able to get the head out, and it will be like that last puppy."

With the cow taken care of, I started washing up. I exhausted my water supply in the truck, and my arm still smelled.

"Maybe you should come in the house and wash again," Dave suggested.

I was quick to take him up on that. I scrubbed and scrubbed on my arm before I felt comfortable putting my shirt back on.

The following day, the view from Dave's place was eerie. Sweet Home was covered with a dense layer of fog. Standing beside the barn and looking out over where you knew the town was, it looked like you were looking out the window a jetliner at twenty-thousand feet: nothing but a layer of clouds.

The cow was in the chute, and it was an easy trip. I scrubbed up the cow and ran an arm in, and there was the head. She probably would have delivered it if I had given her a little more time. Removing it was no problem. I put some more antibiotics into the uterus, and the cow was good to go.

In most cases, that would be the end of the story. But when I stepped into the shower on Saturday morning, I noticed that I had tiny pustules at every hair shaft on my left arm. Had I lived by myself, I would have taken antibiotics off the shelf at the clinic. But Sandy would not hear of that, so it's off to the doctor for me.

On Saturday morning, I have a little trouble convincing Dr. Toffler that I should be looked at today rather than Monday. I think he relented more out of professional courtesy than any actual concern for my arm.

Dr. Toffler looked at my left arm carefully. The pustules ran almost to my shoulder. They were small pustules, and there was no actual discomfort.

"If you want my opinion," I said—I always gave the MDs my opinion; "I think I just need some antibiotics and a few good scrubs."

"I can't figure this out," Dr. Toffler says. "This arm looks just like someone who was in a real dirty hot tub. But the rest of you looks fine."

"Vagina, Doctor, this arm was in a real dirty vagina," I said.

Dr. Toffler shook his head, "What you guys go through, I will never understand."

Ruth and the Goose

"Doctor doesn't generally work on birds," Sandy said to the lady on the phone. "He does make exceptions at times when it is a farm bird and not a pet. I hear you call this goose a pet."

"I only called it a pet because that is how my husband treats it," Sharon replied. "It is the only goose we have. It lives in the barnyard and herds the chickens around all day. He acts like he is the chicken leader."

"Let me go ask the doctor before you come all the way from Brownsville to have him say no," Sandy said as she laid the receiver down.

"I am talking with a lady with a pet goose. It lives in the barnyard. It has a large laceration on its chest. She is wondering if you will take care of it?" Sandy asked.

"If she understands that I treat farm birds like food animals, not like pets, I will take care of it," I said.

Sandy scheduled the appointment, and everyone waited in anticipation for the arrival of the barnyard goose.

When Sharon arrived, the parking lot was packed, and the clinic reception area had no room. It was filled with clients and their pets. I was busy in the exam room, but Sandy popped in and said, "You have to come to look at this."

I stepped out front, and everyone in the waiting room was standing and watching Sharon leading the goose down the street and across the parking lot. She had a twine tied around the goose's neck, and the goose was waddling along like a dog on a leash.

As soon as the goose came through the door, chaos erupted in the reception room. The goose spread his wings and charged at the German shepherd pup. The pup was trying to crawl under the owner's chair to escape the charge. The cat in its carrier on Rosemary's lap was puffed up and hissing at the goose.

Ruth was quick to lead Sharon and her goose back to the surgery room, the only unoccupied space at the time.

When I finally had a few minutes to look at Timmy, the goose, everything in the clinic had settled down. Timmy had a long laceration on the right side of his breast. It was through the skin and extended into the muscle about a half an inch deep.

"Wow, how did this happen?" I asked.

"We have no idea. My husband noticed it when he was feeding the chickens this morning," Sharon said.

"We need to get Timmy under an anesthetic and clean up this wound and close it. Things should go well. Birds have a high body temperature, so superficial infections are not common following wound closure. He will just have a bare patch on his chest for a time. We'll do this right away. I'll have to work him in between patients, and he will have to stay until he recovers from anesthesia. Still, we should be able to send him home early this afternoon."

As Sharon gave Timmy a kiss on his beak, I drew up a dose of ketamine for anesthesia to give as soon as she left.

"I am going to give him an injection of ketamine. This should allow us time to close the wound and have him wake up pretty quickly," I said to Ruth.

"How quick is this going to take effect?" Ruth asked.

"It will take a few minutes. I am going to finish up in the exam room, and then I will be back. It should only take a few minutes to close this wound."

With that, I left Ruth, a short, petite gal, holding a large goose on the surgery table.

I hurried through the vaccination on Rosemary's cat, Whiskers.

"Are you going to be able to take care of that poor goose?" Rosemary asked as we returned Whiskers to his kennel.

"Oh yes, he should be asleep shortly."

All of a sudden, there was a terrible ruckus coming from the surgery room. Timmy was squawking, and we could hear his wings flapping.

"Excuse me, Rosemary. Sandy can check you out. I think Ruth needs a hand."

I rushed to the surgery room. There was Ruth, desperately trying to hold onto Timmy. Timmy was flapping his wings wildly and squawking at the same time. I quickly grabbed him and got his wings under control. Ruth and I held him for a moment, and he drifted off into a deep slumber.

"What caused that?" Ruth asked. "He was fine and then just sort of exploded."

"Just an excitement phase of anesthesia," I said. "It is common with all anesthetics. We just don't see it because what we generally use has such a rapid induction. I haven't seen it with ketamine before, but then, how many geese have we had in this surgery room."

With Timmy under anesthesia, we plucked the feathers around the wound and scrubbed the area with Betadine surgical scrub. After cleaning the wound with a vigorous flush, I sutured the heavy fascia covering the muscle layer with a continuous suture of Dexon, then closed the skin with a buried subcuticular suture, also with Dexon.

With Timmy in a kennel to wake up, we thought the day's excitement was over. That was until the girls were discharging one of the morning surgery cases. The young dog freaked out when he was led past the kennel with a goose flopping about a little. That just made the pup jump about a bit.

***

"Sharon, Timmy is all fixed up," I said as Sharon returned to retrieve Timmy. "He is going to have a bald patch on his chest until he grows some new feathers, but that shouldn't bother him much."

"No, I don't think he will care," Sharon said. "Can he walk?"

"Yes, he is wide awake. We had a little struggle with him as he was going to sleep, but he recovered with no problems. He can walk out of here now. And it is probably a good time since there are no dogs."

"He doesn't like most dogs, and he is pretty protective of his barnyard. He sends our little house dog packing every time he strays too close."

Sharon tied a twine around Timmy's neck, and he hopped out of the kennel. He waddled out our door on his leash, like he knew where he was heading.

Cows Never Eat the Stuff

Here I was again, standing over a dead cow, in the middle of a pasture filled with tansy, listening to a rancher explain to me that cows never touch the stuff. As I stood there sharpening my necropsy knife, I thought about my long history with tansy.

My first recollection of awareness of tansy was at a family picnic at Tom and Kathryn Lawson's ranch, on the top of Catching Creek Mountain, out of Myrtle Point. The year was 1950, the plant was just then starting to show up on the high ridges of Coos County. We knew it killed cows and horses, but I do not recall seeing a loss. My oldest brother had a summer job the following year, working for Coos County spraying tansy, mainly on the high ridges in the county. It did not take long for it to spread to the valleys. By the time I was ten, pulling tansy was a standard summer chore for almost any farm kid in the county.

At least this cow was dead. In the 1970s, the diagnosis was challenging in a live animal before it was near death. Blood work could show liver failure, but that was not specific to tansy toxicity.

I was always amazed at how these guys could be in such a state of denial. They wanted an answer to the death, but one that fits their opinion that cows would never eat the stuff.

It only takes me a few minutes to open this cow up. I slit the skin down the ventral midline and reflect the hide up to her back. Elevating the legs and freeing them of their muscle attachment, I flip both legs and the skin to lie out over her back. Then I open the belly and ribs, reflecting them back, so I now have the cow opened for view.

"So, I want you to look at this, Tom," I say as I start to point out the visible signs of liver failure. "This belly shows all the signs of liver failure. The yellowish discoloration to the tissues, the severe accumulation of fluid in the belly, the chest is normal, and the liver is swollen and pale yellow in color."

"Okay, I can see liver failure," Tom says. "But there has to be a lot of things that cause liver failure. How can you be so sure it is tansy?"

"Well, an old veterinarian, Dr. Pierson, who I respect very much, always said: "When you are in a barn and hear hoof steps, you look for a horse, not a zebra."

"I guess I don't know what that means," Tom said.

"That means you rule out the obvious diagnosis before you go off in some unrelated direction, trying to prove a once-in-a-lifetime diagnosis. In my mind, when I stand in a field filled with tansy, looking at a cow with liver failure, the diagnosis is tansy toxicity. That diagnosis stands until I prove that it is something else. Now let me get a piece of this liver and show you the insides."

I slice off a large section of the liver. The very sharp knife almost vibrates as it is pulled through the dense liver tissue. I lay this piece of liver on the cow's hip for a makeshift table.

"I can send a piece of this into the lab, and the pathologist will give us a confirmed diagnosis," I say. "Tansy toxicity has a very characteristic appearance under the microscope. First, I want you to look at the cut surface of this liver. Think about the liver you see in the store and compare it to this liver. This liver is swollen with rounded edges, not dense with sharp edges, pale yellow in color rather than deep red, and this cut surface has the appearance of nutmeg, not a consistent deep red appearance."

I handed Tom the knife. "I want you to drag this knife through this liver. You watched how sharp this knife is when I opened this cow. I want you to feel how it almost vibrates as it cuts through this chunk of liver."

Tom takes the knife and makes a slice in the liver. "It almost feels like it is cutting steel wool."

"That is maybe a good analogy," I say.

"Okay, Doc, you have presented me a pretty good case," Tom says. "And I guess we are going to send a piece of this to the lab, just to be sure. Why is it then that we don't have a bunch of cows lying here dead?'

"You are a little bit correct, Tom, when you say the cows never eat the stuff," I say. "Most cows will avoid it most of the time. It is most dangerous in the hay and also after it's sprayed. The plant takes up a lot of sugars as it wilts after being sprayed. Cows will find it acceptable for a week or two as it dies. Also, some cows, and some horses, will develop a liking for the stuff, and they will seek it out."

"So you are saying we are both right," Tom says.

"Only sort of," I say. "This cow didn't eat a bunch of tansy yesterday and then died. She could have eaten a toxic dose months ago. You are just lucky that you found her dead and not sick. Making a diagnosis in a cow getting ready to die from tansy is difficult and expensive. Sometimes I will make several visits before ending up doing a liver biopsy. There is a lot of frustration in treating a cow with tansy toxicity. But time always tells us, all of these cows die. I looked at a dead cow once and her 3-day old calf, who was also dead. Both of them died of tansy toxicity. The cow is easy to understand. The calf is a little more of a question. It is doubtful it could have eaten enough tansy to be a problem. There is some evidence that the toxic alkaloids are passed in the milk, but probably not in a dangerous concentration. That leaves the placenta; this calf probably received a toxic dose from its mother through the placenta."

"What do I have to do now to get control of this stuff?" Tom asked.

"It is tough, and it is not going to happen overnight," I said. "Maybe not even in a year. This stuff is too high to benefit from spraying. I would mow it down and either compost it or burn it if you can. Then next spring, you need to spray the pasture. Keep the cows off the pasture after spraying for 2 - 3 weeks. Then next summer, pull any plants that make it through all of that. Probably most important, get all your neighbors to do the same. And talk to the County Extension Agent. They promise that a caterpillar is coming that will eat the stuff. I haven't seen any yet."

"What do I need to do with this carcass?" Tom asked. "Is it toxic? I mean, if my dog gets out here and eats on this, is it going to kill him?"

"Now, that is an interesting question," I said. "And I don't have an answer for that one. I will definitely check the books, but I am not sure anything is written about the toxicity of the tissues. I doubt if there is a problem, but I don't know. I would call the rendering company, or I would bury it."

Don't Die on Me Now

I could hear the old ewe breathing as I approached her in the open pasture. I grabbed her by her long wool, and she made no attempt to move. This summer heat must be unbearable for her. It didn't look like she had been sheared in a couple of years.

I parted the wool on her chest and held my stethoscope against the bare chest wall. I moved to a couple of spots. The air rushing in and out sounded like a freight train. This was severe bronchopneumonia. This old ewe was going die.

I walked over to the blueberry patch where the owner was working.

"This old ewe has severe pneumonia," I said. "I can treat her with some antibiotics, but my guess is she is going to die."

"How long do you think the old girl has to live?" Jim asked.

"Your guess is as good as mine. You maybe have heard the old saying, sheep are born looking for a place to die. I think if she doesn't die tonight, she'll die in the next couple of days."

"Don't you treat pneumonia?" Jim asked, hoping I would give him some optimism.

"Sure, we treat pneumonia. But when it involves the entire lung field on both sides of the chest, there is little chance that treatment is going to do any good."

"She's my wife's pet. I would like to try to save her if that is possible."

"I have a new antibiotic. It is the best one on the market right now. The problem, it's a little expensive. But we could give it a try if you like. I can give her an IV injection now, and if she is alive tomorrow, you can pick up some more to give in the muscle."

"Yes, I think I would like to try that if you think it will work," Jim said.

"I didn't say I think it would work. I think this ewe is going to die. But if you want to treat her, I think her only chance is to use the best drug available."

"What would you do if she was yours?" Jim asked.

"I don't usually answer that question, but if I had an old ewe that sounded like she sounds, I would put her to sleep."

"I think my wife would want to at least try to save her. Let's go ahead and give her an injection today, and I will check with you tomorrow."

I went back to the truck and filled a syringe with an antibiotic injection along with some dexamethasone.

When I got back to the ewe, she hadn't moved from where I had first looked at her. I used scissors to trim some wool away from her jugular vein. I placed a needle in the jugular, attached the antibiotic syringe to the needle, and slowly gave the injection.

Immediately, the old ewe sneezed a couple of times and shook her head. Then she fell forward, banging her nose into the ground, almost lifeless.

"Dammit, don't die on me now," I said to the ewe.

I had only seen one other anaphylactic reaction. That was in a horse when given a penicillin injection while I was in school. That horse reared and went over backward. The horse was dead when he hit the ground.

I looked at the truck. It was probably forty or fifty yards away. If I ran, I could get the epinephrine and be back in fifteen or twenty seconds. I took off as fast as I could across the rough ground.

I grabbed the bottle of epinephrine from the refrigerator and a bottle of sterile water. Then I headed back to ewe at a run.

Epinephrine was one of the drugs that I had to have but never used. I always bought a new set of bottles every year to make sure it was not outdated.

I got back to the ewe and dropped to my knees. I drew a dose of epinephrine into the syringe and diluted it with sterile water. I gave this dose in the jugular vein. It had been more than the twenty seconds I had initially calculated but less than a minute.

The ewe blinked, turned her head, and looked at me. Maybe she would live to see the moon this evening.

She rolled up on her sternum and stood up. That was more activity than I had seen out of her. Probably stimulated from the epinephrine.

I gave her a hefty dose of dexamethasone. Probably not the best practice in treating pneumonia. But in this case, she needs all the help she can get.

"Jim, I had a little excitement with that injection," I said. "The ewe had an allergic reaction. I was able to reverse it with some epinephrine, but it was a close call."

"That's good. Maybe we can have another miracle with the antibiotic," Jim said.

"It'll take a miracle," I said. "Don't be surprised if you find her dead in the morning."

"You don't provide a lot of hope, Doc," Jim said.

"I provide hope when there is hope to give. Otherwise, I provide reality. There is some danger in that reality. Telling somebody that a patient is going to die is one of the riskiest things I do. If she lives, you will be telling me about her for the next twenty years."

"I'll call in the morning, Doc," Jim said.

***

The phone rang early in the morning. It was Jim.

"Doc, the old ewe is dead," Jim said. "Do you think that reaction had anything to do with her death?"

"Jim, it probably didn't do her any good, but the epinephrine gave her more energy than anything else. But remember what I told you before I treated her."

"Yes, I know, you told me she was going to die," Jim said.

Meat is Life

"What would you say if I told you I thought you ate too much red meat?" Dr. Goddard asked.

This was my first appointment with Dr. Goddard. He was trying to get a new style of practice off the ground, and I needed a new primary care doctor.

"I don't think your profession knows squat about nutrition."

"That's sort of blunt," Dr. Goddard says with a surprised look on his face.

"First, eggs are bad. My mother-in-law lived for her eggs in the morning. Her doctors put her through hell for the last few years of her life. Now eggs are fine."

"Okay, I will give you that one," Dr. Goddard said. "But let's get back to the red meat."

"I am not sure you have looked at my file. I am a veterinarian. At heart, I am a cow doctor. Except for my four years in the army, my entire life has been involved with cows. I eat red meat, and that is not a discussion topic for this visit."

"I don't understand how you guys can feel good about caring for animals and then sending them to slaughter," Dr. Goddard said. "The saying, meat is murder, comes to mind."

"I think you are trying to bring this appointment to a close," I said.

"No, I am sorry, I just don't understand," Dr. Goddard said with an apologetic tone.

"Do you want the long story or the short story?"

"I guess I better hear the long story," Dr. Goddard said.

"How far are you removed from the farm? I mean, did your grandfather live on a farm?"

"No, my roots are in the city for a whole lot of generations," Dr. Goddard said.

"So when you drive down the freeway and see hundreds of sheep grazing on the grass seed fields, how many of those sheep do you suppose would be there if they didn't sell lamb chops in the store?"

"I hadn't given it any thought," Dr. Goddard said.

"Not even considering the expense of maintaining a flock of sheep for a year, the labor is considerable. People wouldn't do it for fun."

"I see your point," Dr. Goddard said.

"But you wanted the long story. Those lambs that go to market probably live less than a year. But my profession ensures that their year on earth is good. And we ensure that the meat that reaches the market is the best available in the world," I said.

"A short life is far better than no life. And the market lambs sacrifice themselves to give their mothers and some of their sisters a long life. So I would change your little quip to be more like meat is life," I said.

"And the story goes on. I place at least some of the blame on your profession for the family farm's demise in this country. You guys have been in cahoots with the food industry in your drive to reduce consumption of meat and dairy. Not only the egg issue, but butter is another one. You pushed margarine to replace butter. So you had your patients consuming trans fat instead of butter. Your understanding of cholesterol metabolism at the general practitioner level was way under my training."

"I am not sure I will take the blame for the demise of the family farm," Dr. Goddard said.

"It was a complex issue, but you guys were cheering from the sidelines, at least. And what was the result of that loss? You guys make your recommendations and do your heart surgeries and your drugs, adding a few years onto the life of an old man. With the loss of the family farm, you also lost a whole cluster of farm kids. Farm kids served as stabilizing influencers to their peer groups. Without the farm kids, we have seen drug use spiral out of sight. We lose far more years of life to overdosing than you save in the old men. We lose probably more years yet to the pits of addiction. That all happened at the same time; I don't think you can convince me there is not a correlation."

"You give me something to think about," Dr. Goddard said. "And I guess you are not much interested in my spiel on red meat."

"No, but I want you to know, every time I throw a steak on the grill, I give silent thanks to the animal who provided it. And in a couple of seconds, many animals flash through my mind. The cows I pulled from a creek or saved from death, the calves who I worked so they could be conceived, the steers in the feedlots, and the ones going through the slaughter process."

"So, after all of that, do you still want me as your doctor?" Dr. Goddard asked.

"When I was playing ball in high school, I was told that it was a good thing when a coach chewed you out. If he didn't think you were worth his time, he wouldn't say anything. You just need to mark that file to not discuss red meat."

Run for Your Life

Joe always wanted Comet checked for one thing or the other. He was waiting for his turn in the exam room with Comet on his lap. Comet was a young whippet. There was not an ounce of fat on his entire body. I could about define every muscle on him, just looking.

"What's up with Comet today?" I asked as Joe placed him carefully on the exam table.

"I have been reading about heartworms, Doc," Joe said. "I thought maybe I better have you check Comet and get him on some medication."

"We have just completed a statewide heartworm survey," I said. "One of the drug companies paid for it. Most of the clinics in the state collected blood samples from 100 dogs. They ran all those samples, and they say we have about a two percent incidence of heartworms in native Oregon dogs."

"That doesn't sound like it is too serious," Joe said.

"Not too serious at this point, but they claim that is pretty standard for how heartworms invade an area. It will be in low numbers for several years, and then all of a sudden, it is a major problem."

"Well, even if it is a low-risk thing, I think I want to get Comet on some medication," Joe said. "You know how I am about him. He means as much to me as any of the kids."

"I know, Joe, you have him with you all the time," I said. "The kids come and go."

"Now, don't you tell my wife that I said that. She would be upset with me," Joe said.

"Let me get a blood sample from Comet, and we will see if we can get him on some medication," I said. "This new drug, Ivermectin, is a little bit of a problem with greyhounds and whippets. But the dose used for heartworm prevention is low enough that it is not an issue."

"Whatever you think, Doc," Joe said.

"The risk of the medication causing a problem is very small, Joe. But then, the risk of infection is also minimal. At this point, where you live out on a hillside with few neighbors, I think it is your call."

"That hillside is one of my concerns," Joe said. "We are getting into quite a coyote problem. They are getting so brave that they come right down into the yard and bother Comet. I don't want him catching anything from them."

Comet tested negative for heartworms, and we started him on a new preventative medication.

"You give him one of these tablets once a month," I said. "Try to give it on the same day of the month, but you have a few days leeway if you forget."

***

It was several months later when Joe returned to the clinic. He was distraught, and his body odor told that he had not bathed in several days.

"Doc, Comet is gone," Joe said as he leaned hard on the counter, tears welled up in his eyes. "Those coyotes ate him, I am sure."

"What happened," I asked?

"Three of those damn coyotes came into the yard and started to attack Comet. Before I could do anything, Comet took off like a shot. You know those whippets can run. The coyotes were right on his tail."

"I doubt that those coyotes could catch Comet. I know of ranchers in Colorado who keep greyhounds to hunt coyotes. Those greyhounds just run them down."

"Maybe a half-hour after they left the yard, the whole pack of coyotes were yipping up a storm. I am sure they got him. And he hasn't been home, and that was four days ago. I don't know what I am going to do without him."

"Joe, you need to go home and take care of yourself. Take a shower and get cleaned up. Maybe have the kids help you build a little memorial in the yard for Comet. Then go out for a good dinner. Comet would want you to have a normal life."

"Yes, you are probably right, Doc," Joe said. "I brought this package of pills back. I only used a few, and maybe you can give them to someone who doesn't have the money to afford them."

"We are not supposed to do that, but we keep a few things in the cabinet, just for such a client."

"That poor man," Sandy said after Joe left. "That dog was his whole life."

"He will be okay," I said. "It will just take a little time and some diversion."

***

It was only a few days later when Joe exploded through the door. Exuberant, he had a smile from ear to ear. He had combed his hair, and he was well dressed.

"Doc, I want to thank you for your advice," Joe said.

"You look happy," I said.

"Let me tell you the story," Joe said. "I went home the other day and tried to take your advice, but I couldn't get myself up to it. I laid around another couple of days. Finally, I looked at the yard, and boy, it needed to be mowed. So I went out and started the lawnmower, mowed the yard, and then I took the boys down to Hoy's Hardware to buy stuff for a memorial. And what do you know, when we got home, there was Comet, sitting in the middle of the driveway, waiting for us. I was so happy I almost ran the pickup into the house."

"That is great news," I said. "I bet that Comet ran so fast and so far that he didn't know the way home. When you started the lawnmower, you probably gave him some bearings on how to get home. I didn't think a coyote could catch him."

"You are probably correct," Joe said. "This time of the year, I mow the lawn at least once a week, maybe twice, if I get bored. So Comet would know the sound for sure."

"I would give you those pills back, but I gave them to an old guy this morning," I said with a smile on my face.

"That's okay. I will gladly buy some more," Joe said.

"No," I laughed. "I will grab them for you. I was pulling your leg a little."

Acknowledgments

How does one acknowledge the multitude of people upon whose shoulders he stands?

The many professors and clinicians at Colorado State University College of Veterinary Medicine who spent years ensuring I was prepared for practice. There were times in my years of practice when I had shortcomings, but there was never a time that I could fault my education for those shortcomings.

Possibly by accident, the US Army provided me the maturity to continue and complete my education. And Don Miller, the friend who never had the opportunity to return home, provided me the inspiration to do the same.

Many clients and their animals, both large and small, provided me the stage to practice my profession and enjoy, first hand, the stories laid out in this memoir.

The friends and family who offered encouragement for me to write. And to continue to write, even when my stories were the mere scribbles of an amateur.

Scott Swanson, owner and editor of The New Era, Sweet Home's weekly newspaper, has provided editorial assistance, advice, and column space in his paper and his recent article on my writing endeavor.

Joan Scofield, a long time Sweet Home resident, for her excellent proofreading.

Karla Davenport, a cousin who has broadcast my books to the world, or at least to the Willamette Valley on KFIR radio, 720 AM.

And lastly, Eva Long, who continues to be most helpful in shepherding this venture to completion—something that may not have happened without her expertise.

Photo Credits

Cover: Idella Maeland/Unsplash
Hunting Dogs for Sale: furkanvari/Pexels
Widow Woman's Ranch: Idella Maeland/Unsplash
On a Thanksgiving Eve: Ehoarn Desmas/Unsplash
Edith and Coco: Julian Hauffe/Unsplash
Under the Old Plum Tree: Ivanna Kykla/Pexels
Can We Eat Her?: Alesia Kozik/Pexels
Old Three Toes: David Nieto/Unsplash
Too Many Legs: Klaus Hollederer/Pexels
Cookie's Litter: Brenda Timmermans/Pexels
Long Road Home for Tramp: Gabriel Gheorghe/Unsplash
One More Pregnancy: Harry Cunningham/Unsplash
Polyradiculoneuritis: Guillaume Bourdages/Unsplash
A Perfect Delivery: Helder Sato/Pexels
The Shock of it All: John Marble
Ali: Басмат Анна/Unsplash
The Upgrade: Manny Becerra/Unsplash
Granny's Instructions: Luke Besley/Unsplash
Peanut Digger: Anna Kimbell/Unsplash
Toby's Sore Eye: Echo Grid/Unsplash, D. E. Larsen, DVM
Uterine Twist, Which Way Do We Turn?: Oriel Frankie Ashcroft/Pexels
One Bite Deserves Another: Ivan Vershinin/Pexels
The Perfect Shot: Elina Sazonova/Pexels
A Bear in the Backyard: John Thomas/Unsplash
All the Better to See You With: EVG Culture/Pexels
The Thomas Splint: Product School/Unsplash
TheTurpentine Compress: Hemant Gupta/Pexels

A Hasty Exam: Stephen Leonardi/Unsplash
A Lesson Well Learned: Anthony Beck/Pexels
A Surprise Visit: Dan Meyers/Unsplash
Agroceryosis: Mohau Mannathoko/Unsplash
Back to Her Old Self: Myicahel Tamburini/Pexels
Benefits of Experience: Nastia/Pexels
Rosebud's Wire: Stephen Wheeler/Unsplash.
Choose Your Surgeon with Care: Alice Castro/Pexels
Colleagues: Anastasia Shuraeva/Pexels
Don't Put Her in the Barn: Annie Spratt/Unsplash
All Hell Broke Loose: T. J. Checketts/Pexels
Driving Blind: Carlos Esteves/Unsplash
Elk Delivery: Frank McCubbins
Fleeing the Flea: Liam Ortiz/Pexels
Harry and Buffy: Annie Spratt/Unsplash
The Berserk Mule: Julissa Helmuth/Pexels
Hot Tub Skin Infection: Bob Ward/Pexels
Ruth and the Goose: Alexas Fotos/Pexels
Cows Never Eat the Stuff: D. E. Larsen, DVM
Don't Die on Me Now: Valeriu Bondarenco/Unsplash
Meat is Life: Julia Volk/Pexels
Run for Your Life: Mitchell Orr/Unsplash

About the Author

Dr. David Larsen grew up on a farm in the Coquille River Valley of southwestern Oregon. Animals and their care have been a part of life from the very beginning.

Veterinary medicine was always his radar but it took four years in the Army to provide the maturity for him to complete his education.

First graduating from Oregon State University with a degree in zoology, he then attended Colorado State University, receiving a degree of Doctor of Veterinary Medicine in 1975 at the age of thirty.

With a growing family, he moved first to Enumclaw, Washington, where he practiced for a year and a half. Then he moved to Sweet Home Oregon where he started Sweet Home Veterinary Clinic. He was in practice for over forty years at Sweet Home Veterinary Clinic.

Today he spends his time with his family, writing and doing a little fishing. He and Sandy travel when they can. There is much yet to see in Oregon and the rest of this country.

His first book in the Memoirs of a Country Vet series was *The Last Cow in the Chute & other stories*. Its success and favorable reviews have prompted an accelerated release of the this second book, *Widow Woman's Ranch & other stories*.

Made in the USA
Middletown, DE
08 November 2022